Architectural Structures

Visualizing Load Flow Geometrically

Edmond Saliklis

Routledge
Taylor & Francis Group
NEW YORK AND LONDON

Cover images: Edmond Saliklis. Upper left: Mies van der Rohe's Neue Nationalgalerie, Berlin, Germany; Upper right: Example of the horizontal grid of beams shown in 3D; Lower left: Finding the tributary area associated with column A3 via hand sketch; Lower right: Roof of Pei's Rock and Roll Hall of Fame, Cleveland, OH, as an inclined element

First published 2022
by Routledge
605 Third Avenue, New York, NY 10158

and by Routledge
4 Park Square, Milton Park, Abingdon, Oxon OX14 4RN

Routledge is an imprint of the Taylor & Francis Group, an informa business

Library of Congress Cataloging-in-Publication Data
Names: Saliklis, Edmond, author.
Title: Architectural structures: visualizing load flow geometrically / Edmond Saliklis.
Description: Abingdon, Oxon; New York, NY: Routledge, 2022. | Includes index.
Identifiers: LCCN 2021053249 (print) | LCCN 2021053250 (ebook) | ISBN 9781032018997 (hardback) | ISBN 9781032019000 (paperback) | ISBN 9781003180913 (ebook)
Subjects: LCSH: Structural frames—Mathematical models. | Structural analysis (Engineering)—Mathematics. | Structural analysis (Engineering)—Data processing. | Loads (Mechanics) | Perturbation (Mathematics) | GeoGebra.
Classification: LCC TA643 .S25 2022 (print) | LCC TA643 (ebook) | DDC 624.1/7–dc23/eng/20220201
LC record available at https://lccn.loc.gov/2021053249
LC ebook record available at https://lccn.loc.gov/2021053250

ISBN: 978-1-032-01899-7 (hbk)
ISBN: 978-1-032-01900-0 (pbk)
ISBN: 978-1-003-18091-3 (ebk)

DOI: 10.4324/9781003180913

Typeset in Univers LT
by codeMantra

This book is dedicated to our grandchildren, Matija, Lydija, Montgomery, and June. It is truly a blessing to watch you grow.

Contents

Figures

Equations

■ Equations

Acknowledgments

I have many colleagues at California Polytechnic State University in San Luis Obispo who helped me create this suite of materials. In particular, I am extremely grateful to Professor Graham Archer, who spent many hours with me discussing the details of this method. His critiques and questions were instrumental in making this textbook come to fruition.

My heartfelt appreciation also goes out to my student research assistants, Claudia Zapata-Kraft and Claire Ballard, and to Nathan Lundberg, who was instrumental in clarifying the physical models we created to test these ideas. Finally, I am grateful for the grants I received from the College of Environmental Design at California Polytechnic to support this work.

1 Introduction

Imagine that there was a way of establishing equilibrium of structural elements without any tedious calculations. Imagine there was a way of accurately establishing how much load flows to a support, or how much bending occurs in a particular spot in a beam or frame, all by a quick hand sketch. Imagine that there was a method that could establish load flow in complicated reinforced concrete beams that have changing cross sections. Imagine a method of analysis that is fun, engaging, and truly provides insight into how load flows through a structure.

Your dreams have come true! There is such a technique, and it is the Modern Müller-Breslau Method.

The Modern Müller-Breslau Method is a means of establishing equilibrium of structural elements and assemblies of elements. It does not use traditional algebraic statics, nor does it use graphic statics. Rather, it is an energy method. By perturbing a system slightly, i.e. moving one specific constraint from its original configuration a small amount, the entire structure moves to accommodate that single, small perturbation. As the entire structure moves, so do the external loads applied to the structure move, in accordance with the constraints of the problem. The method hinges on the fact that the work done by the unknown force or moment being sought moving through the perturbation is equal to the work performed by all of the external loads moving through that lofted configuration.

For structural elements, classified as "determinate," the external equilibrating reactions are unambiguous and do not depend on the material properties of the element, nor do they depend on the cross-sectional properties (round, square, and rectangular) of the elements. This is what is meant by "determinate," i.e. the reactions are unambiguously determined. For structural elements classified as "indeterminate," the external equilibrating reactions are not immediately found because the material properties of the element, as well as the cross-sectional shape, do indeed come into play. This is what is meant by "indeterminate," i.e. the deformations and stiffness of the structure must be taken into account when establishing the external equilibrating reactions. The Modern Müller-Breslau Method applies to both determinate and indeterminate structures.

The geometric technique described in this textbook is the Modern Müller-Breslau Method. It has roots in the nineteenth century, but has been expanded in the twenty-first century to include lateral loads as well as gravity loads. While elements of the method are extremely well known, Civil Engineers will recognize the Classical Müller-Breslau Method as the "Influence Line" technique, the modern perturbation technique will be new to readers. The method's application to

DOI: 10.4324/9781003180913-1

two-dimensional grids of beams, to lateral loads, and to indeterminate structures, all of which can be performed on a table napkin sketch, is certainly novel. This new method has not received scholarly attention, with very limited literature devoted to the topic. Detailed examples will be shown using elementary structural elements, for gravity as well as for lateral loading.

Just as *Form and Forces* by Edward Allen and Waclaw Zalewski sparked a revolution in graphic statics, so it is hoped that a similar paradigm shift will occur in the way we teach structures to architecture and engineering students, with the Modern Müller-Breslau Method. While the Classical Müller-Breslau Method is well known to most Civil Engineers around the world, the secrets of this method have been hidden in plain sight for over 150 years. Both the Modern as well as the Classical Müller-Breslau Method can find external equilibrating reactions, in addition to internal axial, shear, and bending moments. These solutions are theoretically exact for any determinate problem subjected to gravity loads only, including complicated grids of horizontal beams. Furthermore, the Modern Müller-Breslau Method can asymptotically approach theoretically correct answers for structures subjected to lateral as well as gravity loading. This textbook will show detailed examples of these methods. Then, this textbook will demonstrate new techniques for analyzing indeterminate structures subjected to gravity loads. The book closes with a few tantalizing areas of projected study such as buckling and plastic analysis.

$$Unknown \cdot \Delta + \sum Force_i \cdot loft_i = 0 \qquad \text{(Equation 1.1)}$$

The method always uses the same fundamental work equation, shown as in Equation 1.1. Note that the signs of the work done by forces (or moments) are indeed important.

In Equation 1.1, the *Unknown* is the force or moment being sought. Δ is a perturbation of the structure, which occurs when employing the method. The symbol Δ is generic; it could be a displacement if seeking an external force or an internal shear, or it could be a rotation if seeking an external equilibrating reaction bending moment or an internal bending moment. Δ is created in the location of the unknown, in some assumed direction. The *Forces* are the externally applied loads to the structure and the *loft* is the travel taken by each Force during the perturbation of the entire structure which was caused by Δ. The main advantage of this technique is that the perturbed shape has absolutely nothing to do with the applied loads. Thus, even for an indeterminate element such as a beam fixed at both ends, the analyst would only be responsible for seeing how the beam would deform due to a given rotation of one end if the moment at that end was sought. This is analogous to the steps taken in the Slope-Deflection Method, where effects are looked at due to individual end degrees of freedom.

STEPS

1. Remove the capability of the structure to carry the *Unknown* you are looking for (reaction or moment).
2. Apply the perturbation Δ in the location of the *Unknown*. In the Classical Müller-Breslau Method this was a unit perturbation; in the Modern Müller-Breslau

Method, this starts out as a visibly large perturbation and is reduced to nearly zero, thereby asymptotically approaching the theoretical answer. Civil Engineers will immediately recognize the Classical Method as the Influence Line problem.

3. Enforce all other boundary conditions.
4. Watch as the external loads are lofted or moved due to the perturbation and measure these lofts. If the method is applied as a hand sketch on determinate structures, the lofts are exactly found by similar triangles. If the method is applied as a hand sketch on indeterminate structures, the lofts are approximated by the sketch which is drawn to some convenient scale.
5. If there is a kink in the structure due to the perturbation, the load must be broken up into sub-portions on either side of the kink.
6. If the structure is determinate, all pieces remain straight after the perturbation.
7. If the structure is indeterminate, *most* pieces are curved after the perturbation and the load must be distributed reasonably smoothly along the curved structural elements.

To introduce the method, consider a statically determinate beam with an overhang, which will be immediately obvious to Civil Engineers. We seek the vertical left reaction, and Figure 1.1 describes the problem. Any reaction symbolized by a triangle will potentially have a horizontal component as well as a vertical component. Any reaction symbolized by a circle will have only a vertical reaction. Loads can be distributed along the length of a structure, as shown in Figure 1.1, or they could be concentrated at one particular point.

Using the previously stated steps, since we seek the left vertical reaction, it is removed. In the Classical Müller-Breslau Method, the perturbation at this point is a unit vertical movement, while in the Modern Müller-Breslau Method, it is a variable movement that follows a circular path. This circular path establishes one major difference between the Classical and the Modern Method. In the Classical Method, the beam is assumed to stretch so that there is no horizontal movement, only vertical. In the Modern Method, only the member being cut and investigated stretches; all other elements remain as rigid body links which move in a circular path and no elements other than the one being investigated change length. If there are no kinks in the perturbed shape, we replace any distributed loads by a single equivalent load at the distribution's centroid. If there are kinks in the perturbed shape, the load must be focused, or concentrated, on either side of the kink, in proportion to the distance the distributed load covers on either side of the kink. However, if there are no distributed external loads, and only concentrated or so-called "point loads", then they do not need to be modified in any way. In Figure 1.1, there are two distinct load distributions, creating two equivalent point loads as shown in Figure 1.2. Figure 1.2 shows the Classical method and Figure 1.3 shows the Modern method of establishing the left vertical equilibrating reaction. Force1 multiplied by Loft1 is

1.1
Example 1 setup

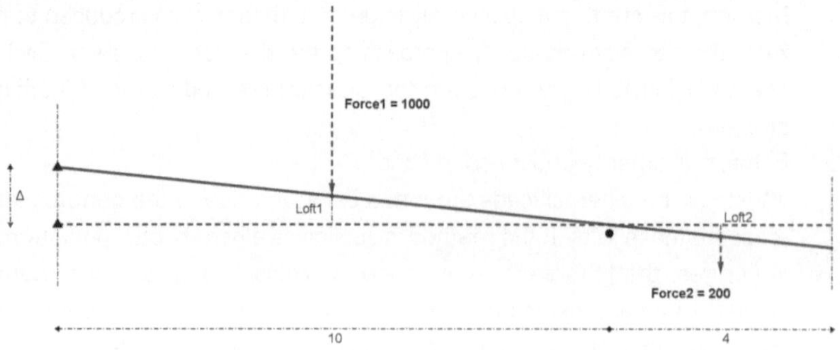

1.2
Classical method for RLeft

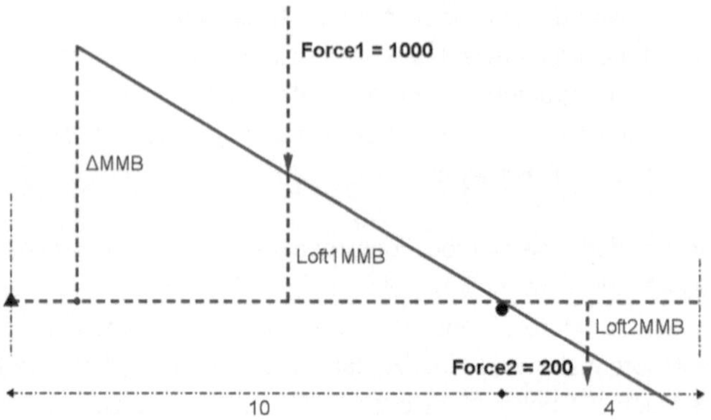

1.3
Modern method for RLeft

negative in Equation 1.1, while Force2 multiplied by Loft2 is positive in Equation 1.1. Note that these work terms are on the left side of Equation 1.1.

For gravity loads on horizontal structures, the Classical Method and the Modern Method are equivalent. One can immediately see how a simple hand sketch can be used in the method since similar triangles establish the lofts effortlessly. Yet, if the structure has lateral loads, or if there is an inclination between the load and the member as in the following example, then the Modern Müller-Breslau Method must be used. Here, hand sketches can be used, but it is much more robust to draw this in a parametric environment on a computer or tablet. In this next example, shown in Figure 1.4, a frame is subjected to a uniformly distributed downward load. The left and right frame members feel bending, but the axial tie member has no bending. Most students would have a difficult time finding the tie force, because of all the differing angles in this problem, but the Modern Method establishes the tie force instantly. The perturbation here is the elongation of the tensile tie, or the difference in length between the perturbed tensile tie and the original tensile tie.

The work associated with each external force in Equation 1.1 (or with each moment) does indeed have a strict sign, and if the direction of the force (or moment) agrees with the direction of the loft, then the work is positive in the fundamental Equation 1.1. In Figure 1.5, it is clear that the work is positive for the downward left and the downward right forces as the tie stretches, but it is less clear what the sign of the tie force should be from a work point of view. It is surely tension, so how is this explained? The easiest way to think about this is to imagine that if a member elongates, the internal tensile force opposes the elongation by "pulling back," thus

1.4
Tied frame setup

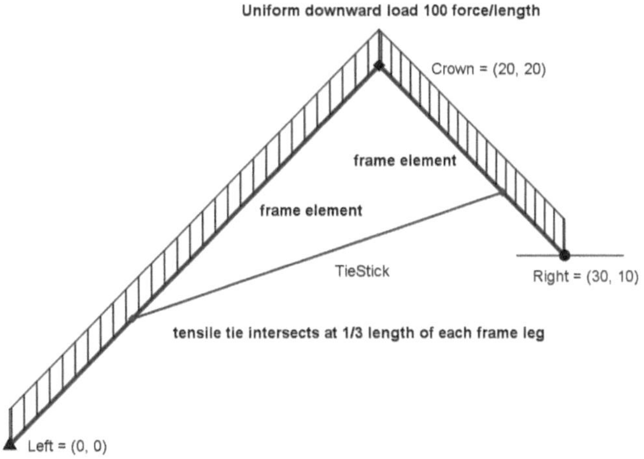

1.5
Perturbed tied frame

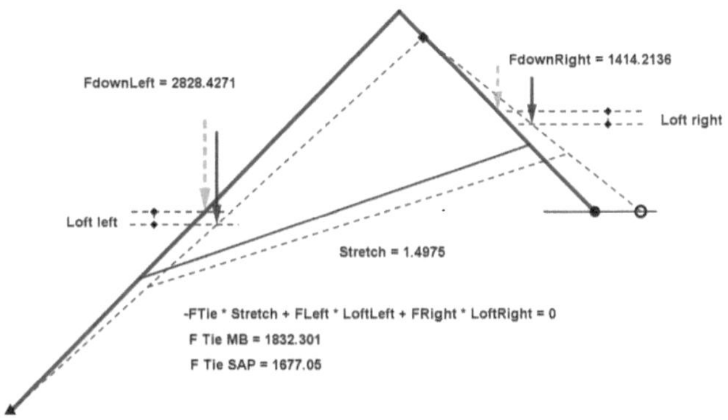

negative work is performed. Conversely, if the member shortens, the internal axial force would "push", again opposing the motion, again creating negative work. This will be further discussed in the chapter devoted to trusses.

Figure 1.6 shows the setup for a determinate, horizontal grid of beams supported by pin-ended columns. Figure 1.7 shows the solution for a single, particular column force, which supports the statically determinate horizontal grid of beams, subjected to gravity loads only. Two primary girders are supported by pin-ended columns, and each girder has a cantilever overhang. A series of pin-ended joists, which are *not shown*, span perpendicular to the primary girders. A uniform load of force/area is applied downward on these joists. One column axial equilibrating force is sought. Figure 1.6 shows the original problem and Figure 1.7 shows the perturbed configuration, where an axial elongation Δ is inserted in line with the *Unknown* of Equation 1.1, which is being sought. Notice that the complexity of the framing of the joists is eliminated; all that is needed is the overall centroid of the downward gravity loads, as there are no kinks in the warped, perturbed surface. The *loft* of Equation 1.1 is the vertical distance between N' and N in Figures 1.7 and 1.6, respectively. It is worth noting that this *loft* could be easily sketched in a notebook, no sophisticated graphics are needed, but the path must be along two mutually perpendicular lines, not diagonal from Δ to N'. One can travel parallel to the girders first, then parallel to the joists. Or one can travel parallel to the joists first, then parallel to the girders, to reach N'.

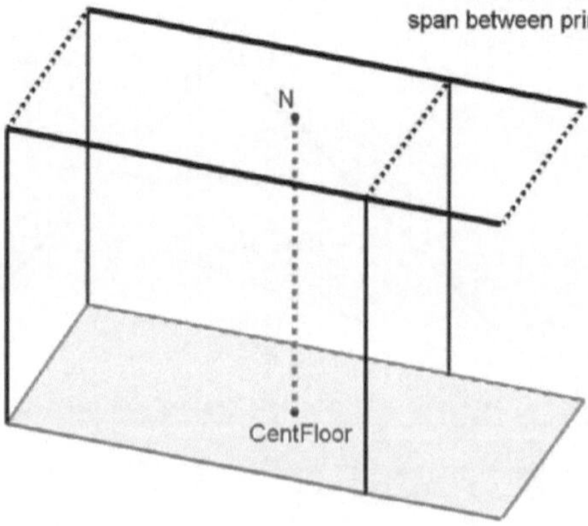

1.6
Two continuous girders support pin-ended joists, joists not shown

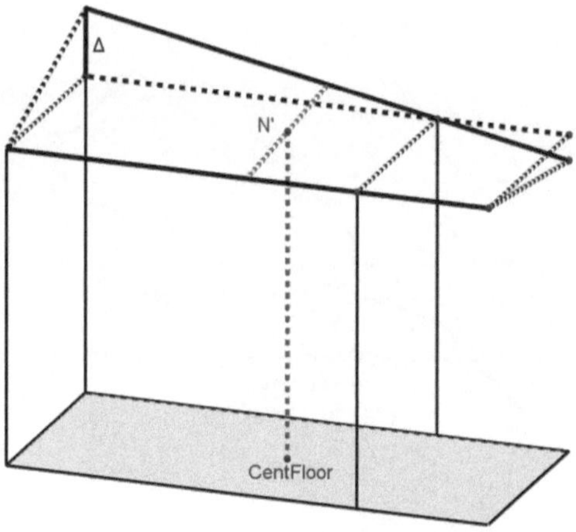

1.7
Classical not modern perturbation to find column equilibrating axial force

In this textbook, for studies of determinate grids of beams will use the Classical Müller-Breslau Method, i.e. elements will stretch as in Figure 1.2, rather than follow circular arc paths as in Figure 1.3.

For the same three-dimensional construction of Figure 1.6, the bending moment at any cut in any beam can be found. Establish the distance "a" from the left end of the beam, and the distance "b" from the right end; then the moment can be found using the method. In Figure 1.8, the Classical Müller-Breslau Method is shown, where the relative angle Δ is assumed to be 1. Most Civil Engineers will remember the construction for finding the height at the cracked, perturbed beam as shown in Equation 1.2.

$$Height\ @\ crack = \frac{a \cdot b}{a+b} \qquad \text{(Equation 1.2)}$$

Figure 1.9 uses a variable angle Δ, and it asymptotically approaches the theoretical answer as Δ gets very small. In Figure 1.9, as will be explained later, the one single element that is studied for the moment at a cut actually stretches on either side of the cut where the moment is sought.

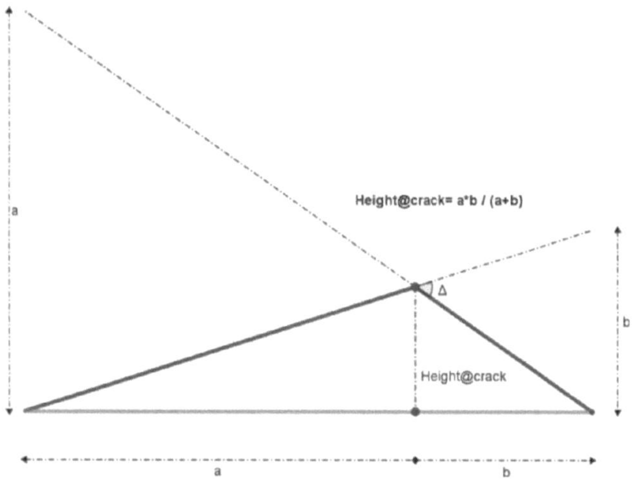

1.8
Classical method of finding moment at some cut

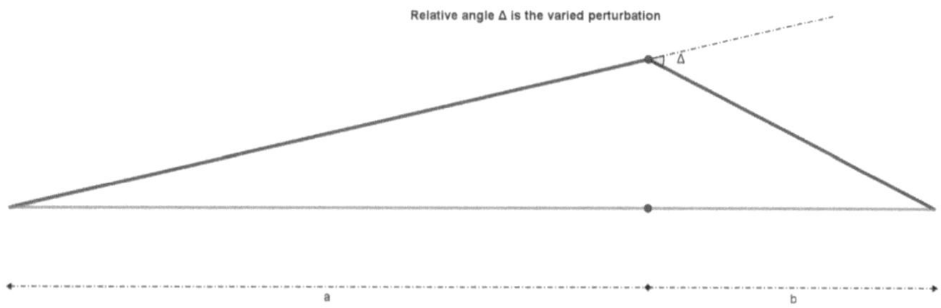

1.9
Modern method of finding moment at some cut

The internal bending moment and the internal shear will be described in full detail in the chapter devoted to beams. Horizontal grids of beams will be further studied as well.

The Müller-Breslau Method can be used to study indeterminate structures. Reinforced concrete beams are a very common typology of indeterminate structures, because the monolithic nature of reinforced concrete ensures the continuity of moments at the ends of beams, thus creating an indeterminate problem. The Müller-Breslau Method agrees well, but not exactly, with the theoretical solution. One advantage the Müller-Breslau Method has in these indeterminate cases is that for a given class of problem, such a single span indeterminate beam, once a carefully determined perturbed shape has been established, that same shape is used again and again for any applied loading. It is worth repeating that the Müller-Breslau Method does not look at deflected shapes induced by loads; it only uses perturbed shapes induced by the movement at the unknown being sought. Another advantage is that it requires the analyst to become comfortable with a limited class of prescribed movements, such as the sole rotation of one end of a beam. This is a very useful skill for structural designers. This is where the intuition of so-called "load flow" in a structure is developed. Whereas traditional algebraic methods can provide an answer, they cannot provide deep insights into how load travels to the equilibrating reactions. The Müller-Breslau Method implicitly requires such an understanding because the perturbed shapes immediately tell the analyst what parts of the structure are "lofted" and thus contribute to the energy of the problem, and what parts of the structure remain inert and contribute nothing.

Figure 1.10 shows an indeterminate beam, subjected to two-point loads, also known as "concentrated loads". The equilibrating moment at the right fixed wall reaction is sought. The details of the construction of the perturbed shape of Figure 1.9 are not described minutely here, but they will be easier to understand after studying the indeterminate examples elsewhere in this textbook. In later chapters, the construction of the perturbed shape of indeterminate structures is explained in more detail. Without a doubt, the perturbed shape is absolutely critical to the success of the method in the indeterminate cases. In the determinate cases, all the perturbed shapes were straight lines, whereas in the indeterminate situation, most lines are curved. If the predicted curves are perfectly accurate, the method will agree with theory. But, since the ease of sketching these shapes must be taken into account, some accuracy is going to be sacrificed to expedite the drawing of the perturbed shapes. Nevertheless, using simple 45° construction angles, the example shown in Figure 1.10 is within 7% of theory.

Since the Müller-Breslau Method requires only the perturbed shape arising from the *Unknown* being sought, and never from the infinitely different applied loads possible, a wide variety of indeterminate problems can be solved. Yes, this does require a development of seeing or intuiting how structures deform due to end translations or rotations, but this is a hugely valuable skill for all structural designers. Experienced readers may question where the "internal energy" is for the perturbed beam of Figure 1.10. The use of Equation 1.1 only looks at "external energy" and it will be wildly incorrect for any structures whose perturbed shapes are grossly incorrect. The idea of a "funicular" or shape that a chain would take under any applied loads is a useful concept here. If the perturbed shape is nearly "funicular" for the unknown being sought, i.e. for a given Δ, then the internal energies disappear in Equation 1.1; they simply cancel each other out. Errors become present if the perturbed shape is not perfectly accurate, which will be the case in indeterminate structures studied later.

For example, a one-story, indeterminate one-bay frame is shown in Figure 1.11. The base of each leg of the frame is pinned creating an indeterminate structure, and the right horizontal thrust reaction is sought. For a given Δ, all other key displacements or "lofts", such as the lateral movement of the top joint, can be found.

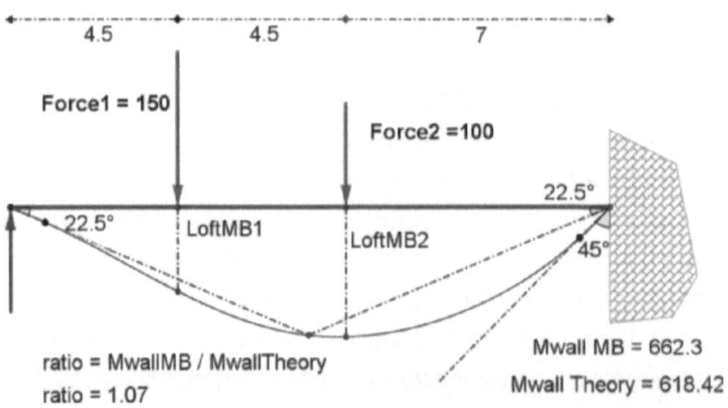

1.10
Propped cantilever, indeterminate beam, right moment reaction found

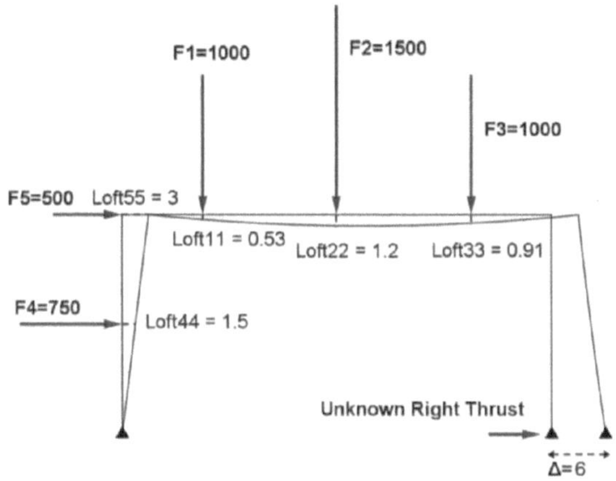

1.11
Indeterminate frame

Exactly how they are found will be described later. For now, notice that all of the lofts are a fraction of the perturbation Δ, which is very large in Figure 1.11. As Δ gets smaller and smaller, the application of Equation 1.1 will provide the correct answer for the unknown force which is being sought, here the right horizontal equilibrating reaction. But this would not be the case if the perturbed shape was wildly incorrect.

The inquisitive reader may question another premise of Equation 1.1, namely where might the ½ be in the traditional work formulation of a slowly increasing load versus displacement curve, where the traditional statement of work is shown in Equation 1.3:

$$\textit{Traditional Work} = \frac{1}{2} \cdot \textit{Force} \cdot \textit{displacement} \qquad \text{(Equation 1.3)}$$

One answer can be found from the well-established field of plastic analysis of frames. Figure 1.12 shows a classic problem that seeks to find the failure load *F*, given a possible collapse mechanism. In the literature surrounding the plastic analysis of frames, the ½ is never present, work is always presented as *Force · displacement* or as *Moment · rotation*

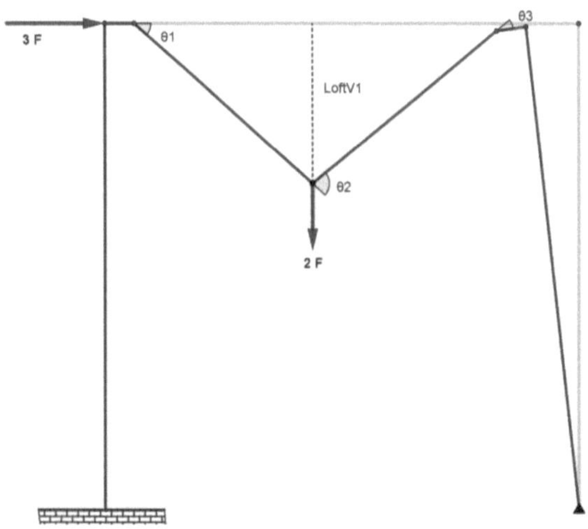

1.12
Plastic analysis of an indeterminate frame

Perhaps an even simpler explanation is that in the Modern Müller-Breslau Method, the external loads that are lofted are moved at full strength for the entire path. Thus, the work is $1 \cdot \textit{Force} \cdot \textit{displacement}$. The ½ would be necessary if the load increased as the deflection increased. Similarly, the unknowns we find in the Modern Müller-Breslau Method are also lofted at full application, not as a steadily increasing force. A third explanation is that if all the forces and the unknown being sought are indeed external, then even if the ½ is present, it cancels in Equation 1.1 because the right-hand side of the equation is zero.

The journey is about to begin! Get ready to sketch, to draw parametrically, and to have fun while learning important concepts surrounding load flow in structures.

2 Determinate Beams

As described in the introductory chapter, there are two classes of structural elements: determinate structures and indeterminate structures. Determinate structures are ones that have unambiguous load flow to the supporting equilibrating reactions. Determinate structures do not take into account the material of the element or the stiffness of the element due to its cross-sectional shape. Indeterminate structures are the ones which require knowledge of the element stiffness, namely the material properties as well as the cross-sectional profile. This chapter will focus on determinate beams. Beams are elements that span over some gap; the span is typically horizontal and gravity loads are always present. But beams might also be inclined as I.M. Pei demonstrated in Figure 2.8 with the Rock and Roll Hall of Fame; thus they would resist downward loads as well as lateral or sideways loads, which may arise from wind or from earthquakes. Whether the element being studied is determinate or indeterminate, the method always uses Equation 1.1 from Chapter 1 of this book, hereafter simply referred to as "Equation 1.1".

The accepted icons or symbols for supporting reactions on such beams are a triangle, and a roller; the triangle means there is a possible horizontal as well as the obvious vertical equilibrating reaction force, whereas the roller represents only a vertical reaction force. In summary, in two dimensions, a determinate beam typically has three supporting reactions, two vertical and one horizontal, but if there are only gravity downward loads on the beam, the "triangle's" horizontal equilibrating force is zero. Yet, some determinate beams will have an internal hinge, and these beams require a fourth equilibrating external reaction. Both of these types will be studied in this chapter.

Figure 2.1 shows a typical setup of a determinate beam problem in a stylized, academic manner of line weight and symbols for supports. In Figure 2.1, we see a pinned support at the left end and a roller support at 10 units of length from the left end. There is an overhang, also known as a "cantilever" that extends 5 units of length to the right of the roller. There are two externally applied downward loads. The first load is at 5 units of length from the left end, and the second load is 12 units of length from the left end. Since there are no horizontal externally applied loads, the horizontal equilibrating reaction at the left end must be zero. Figure 2.1 shows the icons, the "triangle", and the "roller" as well as the possible directions of the equilibrating forces at each of these icons. The reader should note that the "roller" is not like a skateboard wheel; the roller is forced to slide along the surface of the ground; it cannot be lifted up, nor can it move below the surface. The direction of the force will be determined by Equation 1.1. Similarly, the "triangle" icon

DOI: 10.4324/9781003180913-2

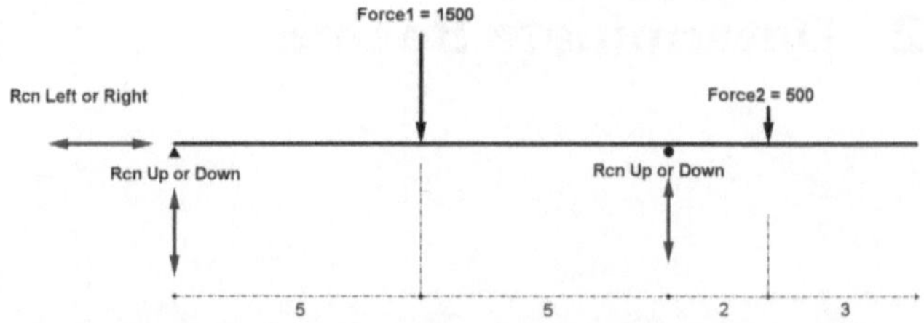

2.1
Setup of a typical determinate beam

represents two forces, one horizontal and one vertical, but the direction of these forces is not known until the application of Equation 1.1. In Figure 2.1, the Left or Right reaction is zero, but in other problems, it might be to the left, or it might be to the right. In all subsequent computer-generated figures in this textbook, these unknown vector arrows will not be shown; rather, the icons of triangle and circle will be used to capture the pin forces and the roller forces, respectively.

Figure 2.2 shows a hand sketch of a horizontal spanning element, similar to the beam element in Figure 2.1, but perhaps more identifiable to the reader. It is a sketch of the front of the Neue Nationalgalerie by Mies van der Rohe. The key feature that the reader should focus on is the tiny gap between the top of the columns and the bottom of the horizontal beam element. There is vertical continuity, but Mies intentionally wanted the roof to appear as if "floating" above the columns. Certainly, sideways or lateral loads cannot be carried by the tiny, nearly imperceptible vertical element between the vertical columns and the horizontal beams.

Figure 2.3 is a hand sketch of a beam being hoisted up during construction. The reader will immediately see the beam is well designed to carry gravity loads; it is deeper near the supports, tapering much like a tree branch does at its unsupported free end, also known as the cantilevered end. This beam was part of the United Airlines Maintenance Hangar in San Francisco, and it was designed by the architect/engineer Myron Goldsmith. Images captured during construction often express the raw, basic function of structural elements in a way that may be harder to see in the completed building.

We seek the equilibrating left vertical reaction for the beam in Figure 2.1. To find this reaction, perturb the left end of the beam along a circular path as shown in Figure 2.4. The circular path must respect all remaining boundary conditions, also known as support constraints. Here, the only other boundary condition is that the beam cannot move vertically up or down at the roller; thus, the left circular arc is 10 units of length, centered on the right roller. In the perturbed configuration, it is clear that the external loads "go for a ride" and as such they clearly move vertically as well as horizontally. But the work in Equation 1.1 accounts only for lofts that are in the direction of the load. In this problem, only the vertical lofts induce work. The

2.2
Ludwig Mies van der Rohe's Neue Nationalgalerie, Berlin, Germany, 1968

2.3
Myron Goldsmith's beam at the United Airlines Maintenance Hangar, San Francisco, 1958

reader must take note of an important idea shown in Figure 2.4, namely that we are using the Modern Müller-Breslau Method and thus the load in the perturbed position changes position both vertically and horizontally, as the load also follows a circular path through the perturbation. Yet, no work is done by the force through its horizontal movement, since the load is vertical. Only the vertical loft "counts".

Even for an enormous perturbation such as shown in Figure 2.4, the answer is exact from Equation 1.1. Notice that the work done by Force1 is negative as the loft opposes the direction of the force.

$$ReactionLeft \cdot \Delta + (-Force1 \cdot Loft1) + (Force2 \cdot Loft2) = 0 \qquad \text{(Equation 2.1)}$$

$$ReactionLeft = 650 \; units \; of \; force \qquad \text{(Equation 2.2)}$$

The Force2 load on the right overhang actually pulls the left reaction downward, as can be clearly seen in the signs of the Forces and the Lofts. This was hinted at in Figure 2.1, which emphasized that vertical reactions could be up or down, horizontal reactions could be left or right, but horizontal reactions are always zero for gravity only loading of horizontal beams. The reader should review one more time that the line of action of any unknown reaction is known, but the actual direction, up or down, or left or right, must be determined by Equation 1.1. Before the actual

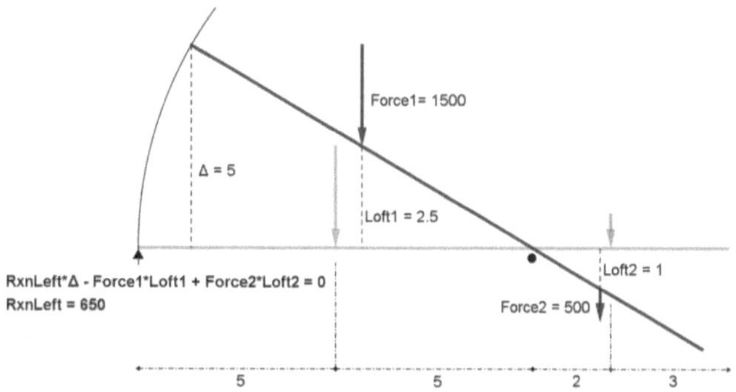

2.4
Solution for the left vertical reaction of a typical determinate beam

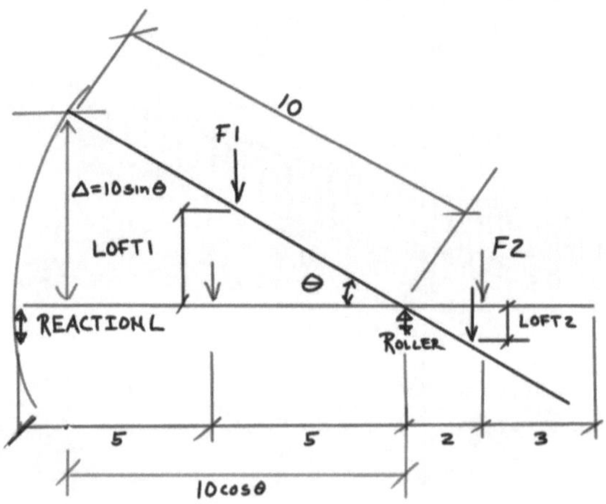

2.5
Solution for the left vertical reaction via hand sketch

programming of such a drawing is described in detail, the reader should take some time to sketch a solution by hand, using large deformations. Not even a straight-edge is needed since similar triangles establish the lofts of all loads. Figure 2.5 demonstrates a typical sketch that should be created by the reader before going on to the next example. The magnitude of the angle θ can be fairly large, 45° is a convenient angle, but the reader should recognize that similar triangles immediately and unambiguously establish the lofts of each load for any angle.

The height at the perturbed tip in Figure 2.5 is $10\sin(\theta)$ and that single height establishes Loft1 and Loft2 via similar triangles. For those who do not remember the rule for similar triangles, Equation 2.3 shows the similar triangle application. Figure 2.6 reinforces those similar triangles for the beam of Figure 2.5, with special attention paid to the similar triangles that are formed between the perturbed beam and the original, unperturbed beam for some arbitrary angle θ.

$$\frac{10\cdot\sin(\theta)}{10\cdot\cos(\theta)}=\frac{Loft1}{5\cdot\cos(\theta)} \qquad \text{(Equation 2.3)}$$

For elements subjected to lateral loads, i.e. horizontal loads, the technique is the same as for gravity loads, but now the loft being captured is in the direction of the load, which for lateral or sideways loads is horizontal, not vertical. Figure 2.7 shows an inclined beam subjected to a single lateral load. The left end is pinned and there will be an equilibrating horizontal reaction at this support, which must be equal to

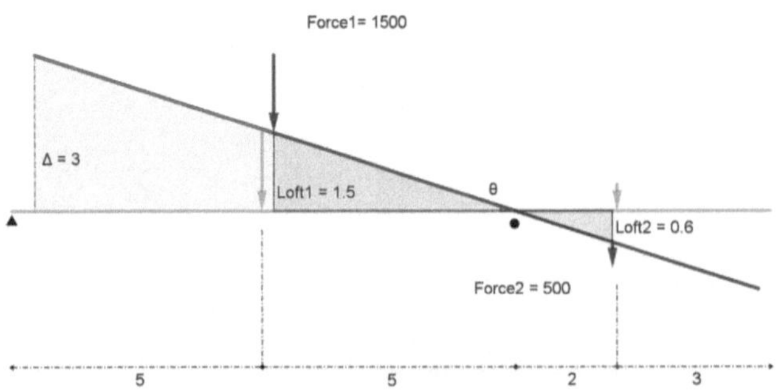

2.6
Similar triangles in detail

2.7
Simply supported inclined beam with one lateral load applied

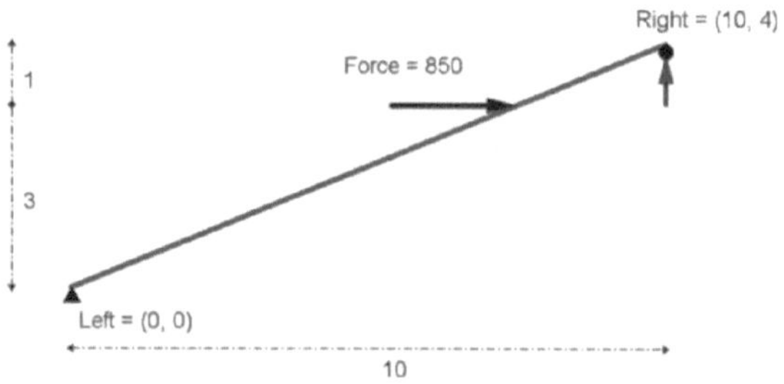

the magnitude of the applied lateral load. There is a roller support at the right end, with only a vertical equilibrating reaction, and there is another vertical equilibrating reaction at the left end. Can the reader speculate which direction that right vertical reaction is directed towards, up or down? In this example, we seek the right vertical reaction at the roller. Furthermore, in this example there is no overhanging portion of the beam; we often call such beams without overhangs as "simply supported".

Before working on the solution to the inclined beam problem, the reader should imagine what architectural applications commonly use inclined members to carry loads. Figure 2.8 shows the inclined roof of I.M. Pei's Rock and Roll Hall of Fame, a dramatic sloped element that resists gravity loads as well as lateral wind loads.

Since the right vertical reaction support of the beam of Figure 2.7 is sought, the Modern Müller-Breslau Method requires the removal of that reaction, a vertical perturbation in the form of a circular linkage path, and a complete respect for all other boundary (or support) conditions, here the pinned left end. Figure 2.9 shows the perturbed inclined beam. As noted earlier, the load itself follows a circular path through the perturbation, thus it has a vertical loft and a horizontal loft. Yet, because the load is horizontal, only the horizontal loft creates work, the vertical loft is not used at all.

As the perturbation gets smaller and smaller, the solution for the unknown right vertical reaction will approach the theoretical answer of 255 units of force. Notice that the work done by the lateral load is negative because the loft opposes the direction of the force, i.e. as Δ goes up, the loft goes left but the load goes right, as shown in Equation 2.4.

2.8
Roof of I.M. Pei's Rock and Roll Hall of Fame, Cleveland, OH, as an inclined element

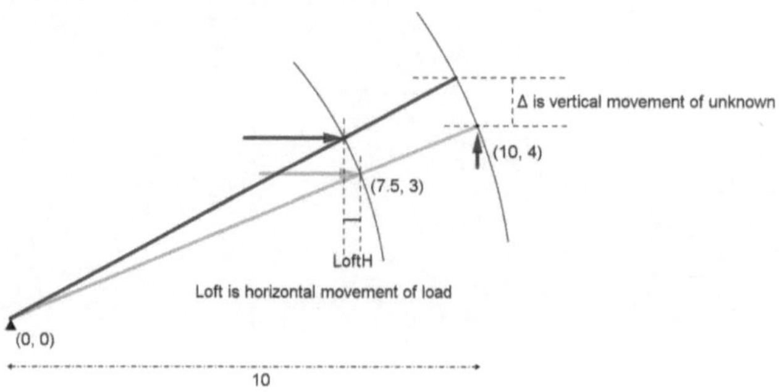

2.9
Establishing Loft and Δ for Equation 1.1

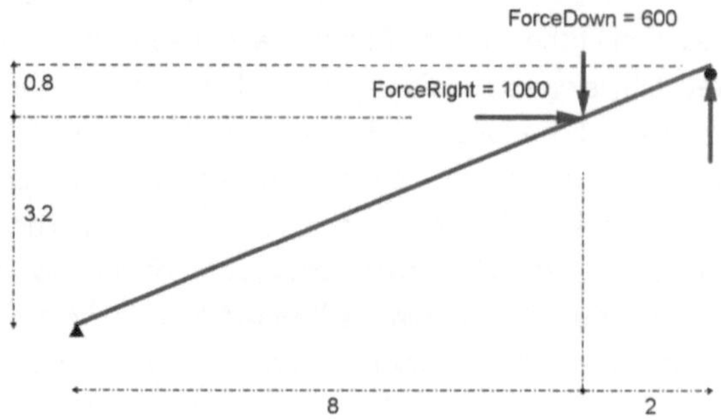

2.10
Vertical and lateral load applied: finding the right vertical reaction

$$Right\ Rcn\ Unknown \cdot \Delta + (-FLateral \cdot LoftHorizontal) = 0 \quad \text{(Equation 2.4)}$$

The next example combines a downward load with a lateral load. It is worth noting that all the problems in this textbook adhere to an important concept in engineering known as "superposition". What superposition means is that we can simply look at one case, and then look at a separate case, and the addition or superposition of the two individual cases results in the combined effect of both. Thus, we could analyze the beam of Figure 2.10 twice, once for gravity load and once for lateral load, but here we will analyze the beam for both loads applied simultaneously. To simplify the problem a bit, both the vertical load and the lateral load are applied to the same point on the beam, but in other situations the point of load application might differ for each load.

In Figure 2.10, we seek the vertical equilibrating reaction force at the right roller. As such, we remove that reaction and we perturb the right end vertically, but along a circular path that respects all boundary conditions. Here, the only other boundary condition is that the beam cannot move from its left end; it is pinned to the ground there.

In Figure 2.11, as before, the right tip of the beam swings in a constrained circular arc, whose radius is the length of the beam. And the point of load application is clearly marked as 8 units horizontally and 3.2 units vertically of length from the left pinned support. But where is the load located on the perturbed beam? Is there a simple way of locating that point in space?

The answer is "yes", but the reader should sketch out a possible means of obtaining that point, preferably a means that is geometric and doesn't require too many calculations. The solution of that small puzzle leads to an important point

2.11
Construction to find the right vertical reaction

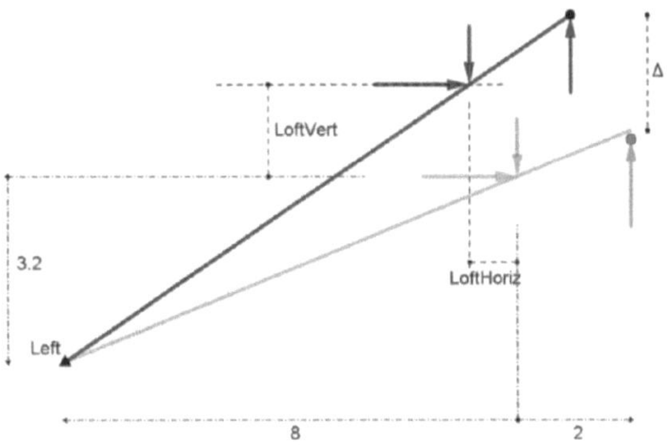

about methods in this textbook, and to a large extent, the methods of structural analysis. Namely, that there are always multiple ways of solving any one given problem. Some methods are direct and elegant, other methods are lengthy and somewhat tedious. One of the joys of solving problems geometrically is that the analyst is free to solve problems in any manner which pleases her or him; some may be more efficient, but many paths will lead to the final answer.

One particularly quick and fairly elegant manner of locating that point on the perturbed beam is to simply draw a circle, centered on the left pinned support, and to pass that circle through the load point on the original, unperturbed beam. Then, the load point will always be on that circular path, regardless of the magnitude of the perturbation. This is demonstrated in Figure 2.12. Notice that no sines or cosines were used in Figure 2.12, although they could have been used. This method of finding loads on the perturbed circular path was used previously in Figure 2.9. Clearly, the free tip, as well as any other point on the rigid, perturbed beam, follows a circular path.

Then, Equation 1.1 is used to find the unknown right vertical reaction. Notice that the work done by both external loads is negative in the construction of Figure 2.9. The work is shown in Equation 2.5.

$$\begin{aligned} &Right\ Rcn\ Unknown \cdot \Delta + (-FLateral \cdot LoftHoriz) + \\ &\qquad (-FDown \cdot LoftVert) = 0 \end{aligned} \quad \text{(Equation 2.5)}$$

2.12
Construction to find the location of load point on the perturbed beam

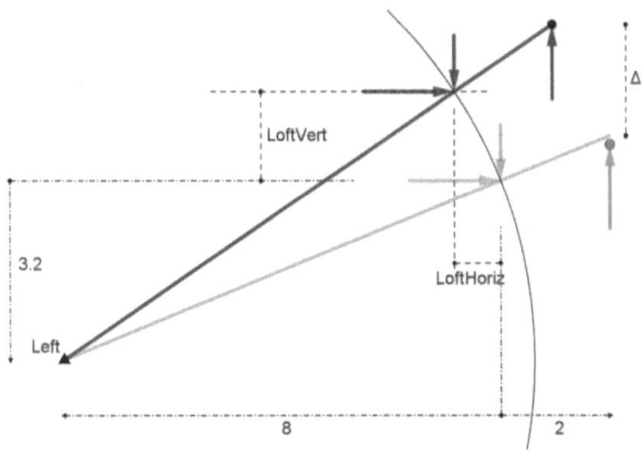

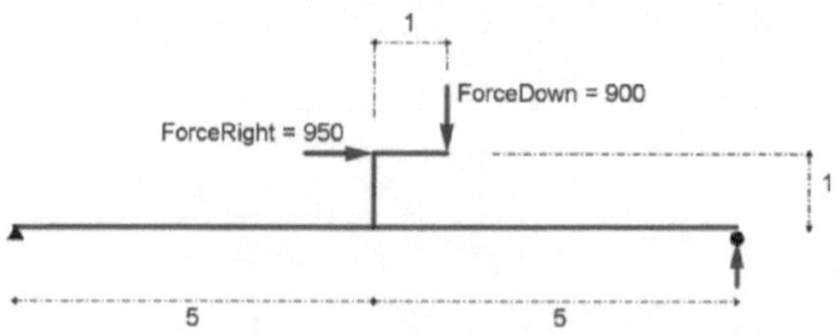

2.13
Setup of an indirectly loaded beam

With a small perturbation Δ, the Right Reaction approaches 800 units of force.

The reader should re-create Figures 2.11 and 2.12 by a hand sketch, before attempting to program the drawing on a computer. The hand sketch will reinforce what items are necessary such as Δ or *LoftHoriz*.

The following problem is slightly academic, in that the structural element is indirectly loaded through an "appendage". However, it begins to demonstrate that the complexity of the problem setup does not add to the complexity of the solution. Equation 1.1 is always used and is always the same, no matter how complicated the problem appears. In Figure 2.13, we seek the right vertical equilibrating reaction of the beam that is indirectly loaded.

The technique is always the same, namely removing the unknown being sought, perturbing it in the direction of the unknown, respecting all remaining support, or boundary, conditions, and following a linkage, or circular path during the perturbation. As before, the reader is encouraged to sketch out the perturbed shape by hand, or perhaps to simply imagine how the perturbed shape would look. Both exercises have great benefits and will aid the reader in understanding the material, prior to actually programming it in a drawing environment.

Before showing the solution, however, Figure 2.14 shows an example of essentially the same system as in Figure 2.13, but in an architectural setting. Architect Neil Denari created a small residence that has the primary beam member supported by columns. Above that beam however is an appendage that is functionally similar to the appendage of Figure 2.13.

Figure 2.15 shows the solution. Notice that the perturbation is enormous. But make no mistake, the loads remain purely vertical and purely horizontal even in the perturbed situation. This is why we seek only the vertical loft for the vertical load and only the horizontal loft for the horizontal load. The method approaches the theoretical answer for very small perturbations. For problems that have only a vertical load, very large perturbations still result in the exact theoretical answer; yet the technique always assumes that the line of action of the applied loads does not change during the perturbation. Figure 2.15 is grossly distorted to clarify the items needed in Equation 1.1.

And the application of Equation 1.1 provides the unknown right reaction, provided that Δ is very small. In this problem, both loads do negative work since their lofts oppose the direction of the applied force. It is worth noting that the unknown right reaction does positive work in Equation 2.6.

$$\mathit{Right\ Rcn\ Unknown}\cdot\Delta+(-\mathit{FLateral}\cdot\mathit{LoftHoriz}) + (-\mathit{FDown}\cdot\mathit{LoftVert})=0 \qquad \text{(Equation 2.6)}$$

2.14
Neil Denari's Alan House, Los Angeles

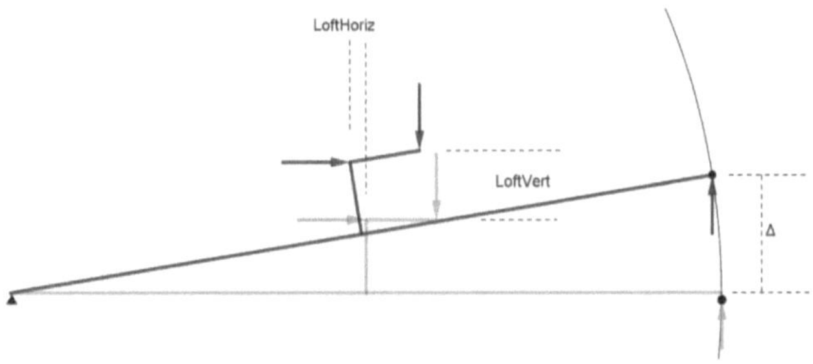

2.15
Solution for the right reaction of an indirectly loaded beam

As the perturbation Δ gets small, the Right Reaction approaches the theoretical value of 635 units of force.

Until now, we have considered only simply supported beams or beams that have an overhang beyond the simple supports. There is another restraint condition however which is extremely important, namely the fixed connection, or the bending moment carrying connection. In everyday terms, one can think of this as the supporting connection that a branch has to a trunk of a tree, the branch "cantilevers" off of the trunk and the branch would indeed transfer bending to the trunk if a person were to sit on the branch. Another everyday image is that of a diving board, the board juts out into space or "cantilevers" out to space, but it is fully restrained from translation vertically, translation horizontally and from rotation at the supporting end. This is what is meant by a fixed connection. Another image that is less obvious perhaps is that of a tall building. A tall building is always a vertical cantilever. It springs forth from the ground and is fixed at its base. We seek the equilibrating reactions at the fixed connection of any cantilever, whether it is horizontal like a tree branch or vertical like a tall building. Consider the horizontal cantilever shown in Figure 2.16.

2.16
Typical cantilever beam

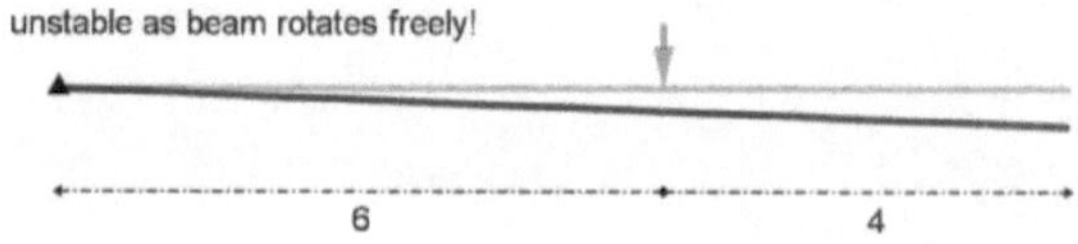

2.17
A pinned cantilever is always unstable

What restraints are needed to keep the cantilever beam in equilibrium? The answer is that three restraints within the "wall" or the support are needed. One is a horizontal reaction, in case there were horizontal or lateral loads applied to the beam. One is a vertical reaction, to counteract the traditional downward vertical loads applied to the beam. But a horizontal reaction and a vertical reaction define a so-called "pinned connection" whose icon we saw previously is the triangle. Why is a pinned connection insufficient in a cantilever beam? Figure 2.17 shows why.

Many horizontal cantilevers exist in architecture; the drama of the "floating mass" is irresistible to architects and a source of concern to structural engineers. Figure 2.18 shows the new Modern Art Museum MO in Vilnius, Lithuania by Daniel Libeskind, whose dramatic cantilever hovers over the entire entrance staircourt.

Figure 2.19 shows qualitatively, not quantitatively, the reaction moment in the wall that prevents spinning. In Figure 2.17, the beam is unstable because the pin at the left end restrains movement, also known as translation, but it cannot restrain spin, or rotation. An engineering term for this spinning effect is known as "moment"; a "moment" is a force multiplied by a distance perpendicular to the line of action of the force. In Figure 2.17, it is perhaps intuitive to the reader that the further the force is from the pin, the larger the spin or moment would be, even if the force does not change magnitude. That distance that causes a force to spin about a point is called a "moment arm"; it is intuitive on a playground teeter-totter, the further a body (force) moves from the fulcrum (the pivot point), the more spin or moment is induced about that pivot point. To equilibrate the beam of Figure 2.17, a reaction that is a moment must be in the wall, counteracting the spin applied by the force at some distance to the wall. This reaction moment is very important

2.18
Daniel Libeskind's cantilever at MO Vilnius, Lithuania, 2018

2.19
Cantilever requires an equilibrating moment at support

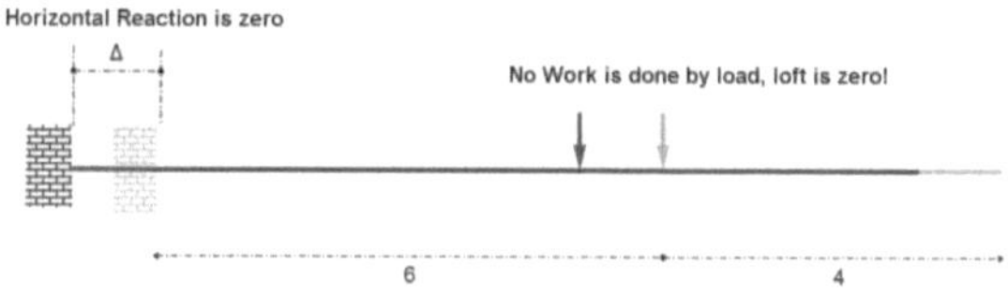

2.20
Beam with no lateral load has zero force horizontal reaction

and it will take some time for the novice reader to understand that concept if this is the first time it has been studied. For students who have some background in engineering or physics, this reactive, equilibrating moment will be quite apparent as a counterclockwise reaction in the wall at the left end of the beam in Figure 2.19, which shows that internal equilibrating reaction moment in the left support.

The Modern Müller-Breslau Method can be used to find all three equilibrating reactions. While the horizontal equilibrating reaction is trivial to find, it is worth doing once for completeness. Figure 2.20 shows why the horizontal equilibrating reaction is zero.

Figure 2.21 shows that the solution for the equilibrating vertical reaction in the wall is simply equal to the summation of the downward forces. Intuition reinforces this finding, in the form of Newton's famous "equal and opposite" reactions.

Yet the equilibrating moment reaction in the wall of Figure 2.19 may cause a bit of puzzlement in the reader's mind. Equation 1.1 is still used, but now the unknown "Δ" is in fact the rotation at the junction of the wall and the beam. This rotation is perturbed from its original pristine condition. Equation 1.1 will always use the symbol Δ as the perturbation associated with the unknown. For unknown forces, Δ will be a movement, also known as a translation. For unknown moments, Δ will be a rotation. Figure 2.22 shows the perturbation needed to establish the equilibrating moment in the left wall. The beam is cracked there, preventing any transfer of moment from the beam to the wall. The perturbation is the angle that the beam makes with a small horizontal segment of the beam embedded in the wall. As the angle changes, so does the loft of any applied load change. Then Equation 1.1 immediately finds the equilibrating moment, when the angle is expressed in radians, not degrees. Note that the units of moments are always *force·length*. It is worth noting yet again that the externally applied downward load has a vertical loft and a horizontal loft. But we only can use the vertical loft, since the load is purely

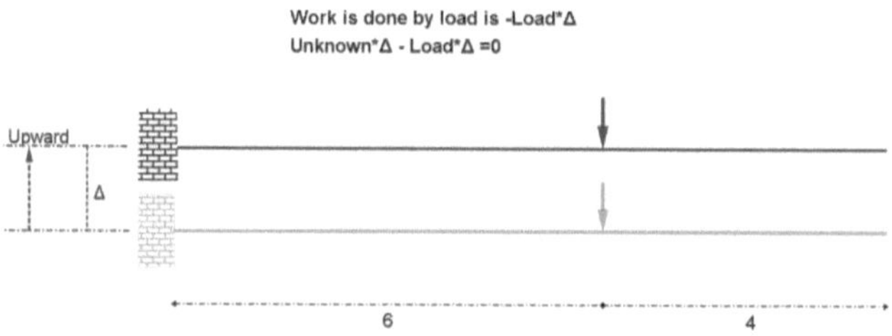

2.21
Beam with one vertical reaction equilibrates all downward load at one point

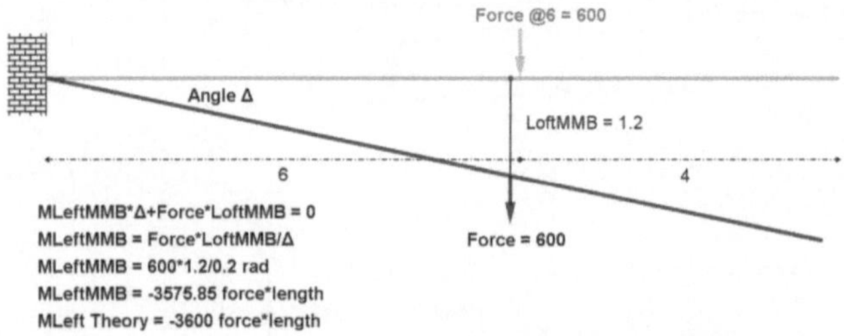

2.22
Establishing the equilibrating reaction moment at the left end

vertical. It remains vertical through the perturbation. Notice the required use of radians, not degrees, in Figure 2.22.

Yet another type of determinate problem is the so-called beam with an "internal hinge". The practicality of such an internal hinge is very real. If a beam needs to be shipped to the site on a truck, the beam has a finite length that can be accommodated by standard trucks. At the site, the beam needs to be connected to a subsequent beam, and that connection might transfer bending if the connection is fully welded. However, if the connection is bolted and not field welded, it may be appropriate to model it as not being able to transfer bending to the next member. This is analogous to a door not being able to transfer bending to the jamb. The door can transfer force, but it is impossible to bend the jamb by moving the door; thus the name "internal hinge" is used, much like a door hinge. Note that a beam with an internal hinge always has four Unknown external reactions in a two-dimensional space, as opposed to the traditional three Unknowns as was shown in any of the beams studied in this chapter. Figure 2.23 shows a beam with an internal hinge 5 units away from the left fixed support. As always there are three equilibrating reactions in the fixed end. There is also a fourth reaction at the roller 12 units from the left support.

Before studying any particular reaction of the beam in Figure 2.23, it may be helpful to the reader to imagine a long beam in an architectural setting, one that requires some kind of hinged connection between two long elements. Figure 2.24 shows a large timber arch with a shear connection, as well as a possible variation of a shear connector in a close up view.

Returning the problem of a beam with an internal hinge, suppose we sought the right reaction for the beam in Figure 2.23. The fundamental technique always remains the same, namely to remove the unknown, perturb the structure to an amount Δ in the direction of the unknown, and respect all other boundary conditions. If a kink is formed in the structure, any uniform loads across the kink must be broken up into equivalent point loads on either side of the kink. This last rule comes into effect in this example as shown in Figure 2.25. Since there are only gravity loads here, the answer is exact, even for an enormous perturbation. But

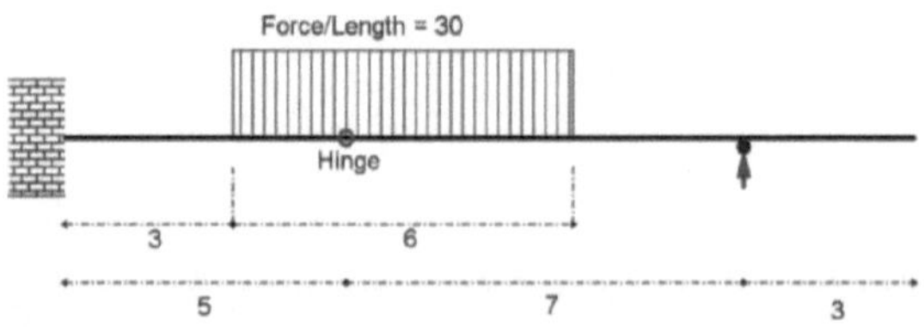

2.23
A determinate beam with an internal hinge

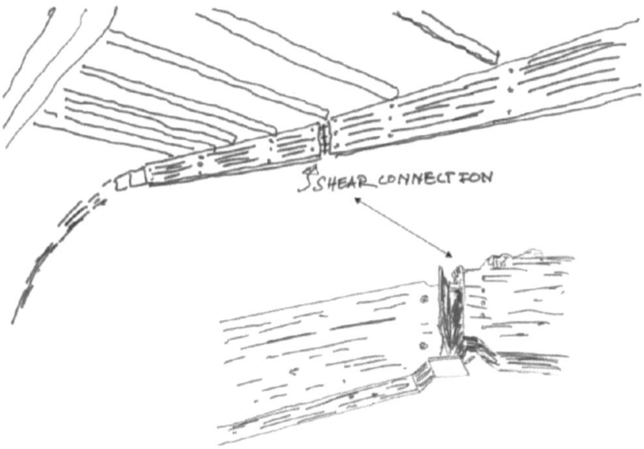

2.24
Connecting two beams with an internal hinge

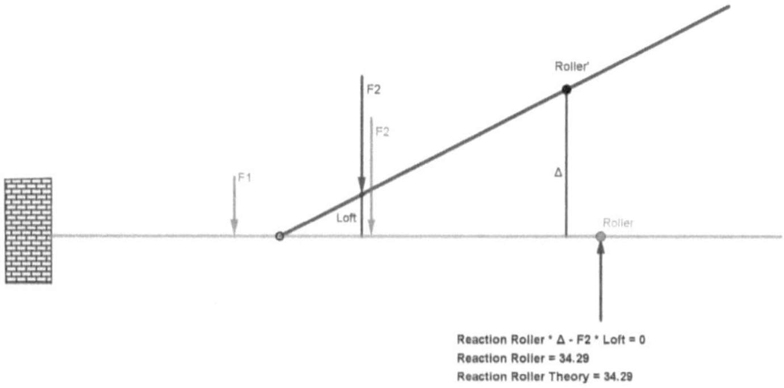

2.25
Solving for the right reaction of the propped cantilever

the most important insight from Figure 2.25 is the fact that only the portion of the distributed load that is to the right of the internal hinge influences the right roller reaction. The novice reader may be surprised by the fact that the portion of the distributed load to the left of the internal hinge does not influence the right reaction. This insight is immediately apparent using the Modern Müller-Breslau Method, as only F2 is lofted, not F1. As the perturbation Δ gets small, the Reaction at the Roller approaches the theoretical value of 34.3 units of force.

Until now, the examples in this chapter have shown the process of obtaining external equilibrating forces and moments. But internal forces and moments also exist inside of structural elements, equilibrating them against applied loads. For the novice reader, this may be a bewildering concept, the idea of forces inside of a structure. Yet common intuition can help understand this phenomenon. Return to the previous thought experiment of the tree branch cantilevering from the tree trunk, the equilibrating reaction of the internal moment was established and stylized at the left end of Figure 2.19. But the entire branch bends, all the way from the tree trunk, until the very end of the branch, where there is no more bending. How could the internal bending moment be found at any point in the branch? The answer, of course, is still Equation 1.1. Halfway along the length of the branch, bending exists. Three quarters of the way from the trunk to the tip, less bending exists. At the free end, no more bending exists.

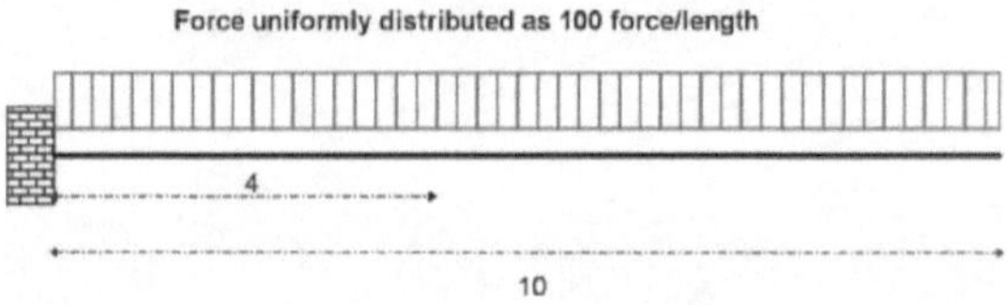

2.26
Seeking internal moment at some cut in the cantilever

Suppose a cantilever beam was 10 units long, fixed at its left end. The internal moment at some cut, say $x = 4$ units of length from the left end is sought. The setup of the problem is shown in Figure 2.26.

Intuition would tell us that, if this were a branch on a tree, the bending at 4 units of length from the support must be less than the bending at the tree trunk support itself. The technique is the same as before, remove the unknown, which is the internal bending moment at $x = 4$, perturb the structure, and since there is a kink in the structure, place the uniformly distributed load proportionally on either side of the kink. Figure 2.27 shows the solution for a very large perturbation. As the perturbation (here the angle Δ) gets small, the answer approaches the theoretical value. Yet again we emphasize that the load has vertical as well as a horizontal loft through the perturbation. But since the load is vertical, we use only the vertical loft. The horizontal loft has no influence. Notice that the moment at the cut, i.e. the Unknown of Equation 1.1, Chapter 1, turns out to be negative. An easy way for the novice reader to check whether the sign of a moment should intuitively be positive or negative is this simple rule. A "happy beam" has the general shape of a smiley face, i.e. concave up is positive and the "sad beam" or concave down, means negative bending. Using this easy to remember rule, the general description of the perturbed beam is "sad" or negative in Figure 2.27.

The calculation of internal forces and internal moments in typical beams is a very practical problem that needs to be studied by everyone interested in building design. The most common and simplest case is the so-called simply supported beam, previously defined as having no overhangs and two vertical reactions and one, typically zero horizontal reaction. Figure 2.28 shows the nearly exact solution for a simply supported beam that is 12 units of length long, subjected to a single downward force at some distance x from the left end, here that distance is 4 units of length. But the internal moment at midspan is sought. Equation 1.1 immediately establishes the magnitude of that moment. Since we seek the internal bending moment which depends on the angle Δ, the answer is asymptotically

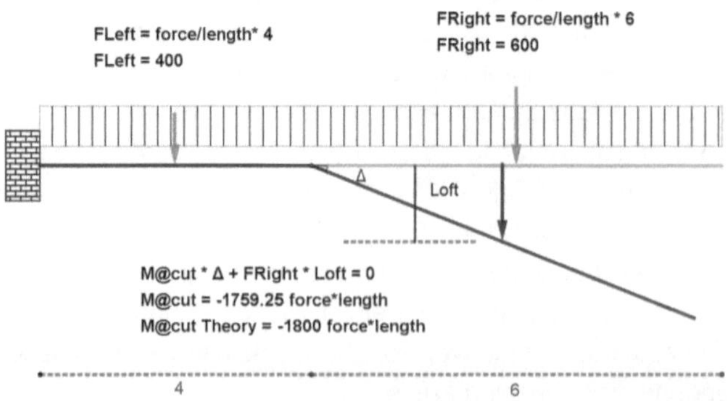

2.27
Finding internal moment at some cut in the cantilever

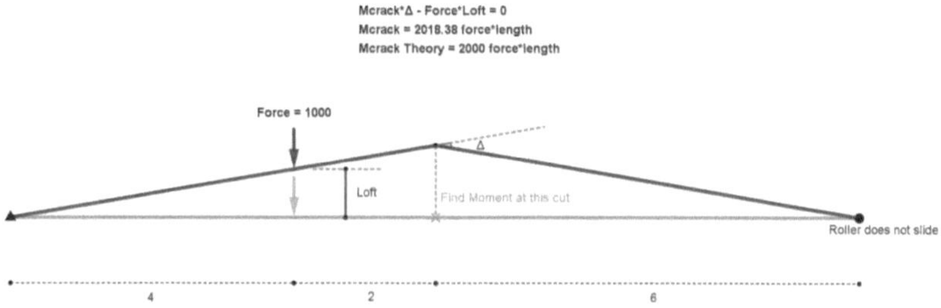

2.28
Finding internal moment at some cut in a simply supported beam

approached. This is the Modern Müller-Breslau Method, not the Classic Method, thus we can reduce Δ till we approach the theoretical answer. Furthermore, a subtle but important idea is demonstrated in Figure 2.28, namely that when seeking an internal moment (or an internal shear discussed later), the single element being investigated stretches on either side of the cut that exposes the moment. All other elements follow circular paths, as will be shown later. That is why "the roller does not slide" in Figure 2.28.

It is very important to establish a few guiding principles that must be maintained in the Modern Müller-Breslau Method.

- External loads do not change orientation through the perturbation, i.e. vertical loads remain vertical, horizontal loads remain horizontal.
- Only movement or "lofts" in the direction of loads are used in Equation 1.1 of Chapter 1.
- Only the member being studied for the Unknown changes the length and shape, all other members never change the length or shape for determinate problems.
- Angles in frames or arches at moment carrying connections never change during the perturbation.
- All boundary conditions must be enforced.
- All elements other than the element with the Unknown remain perfectly straight for determinate problems.

To highlight these guidelines as they apply to the simplest possible case, namely a simply supported beam, consider a simply supported beam subjected to a downward force and to a horizontal force as shown in Figure 2.29. Both of these forces are applied to a single point, somewhere left of midspan. But we seek the internal moment at the midspan of the beam.

Figure 2.30 shows an incorrect construction of the perturbed shape, to highlight a guiding principle previously stated. If we constructed the perturbed shape such that the length to the left of the cut remained at L/2 and the length to the right of the cut remained at L/2, then the right Roller Support would have to slide to accommodate this so-called "linkage" or mechanism. But this is obviously incorrect because this linkage would cause the horizontal load to loft or to move, and that would add a term to Equation 1.1 of Chapter 1 arising from a horizontal load. That would erroneously mean that a horizontal load applied to the horizontal beam would induce bending, and that is certainly wrong.

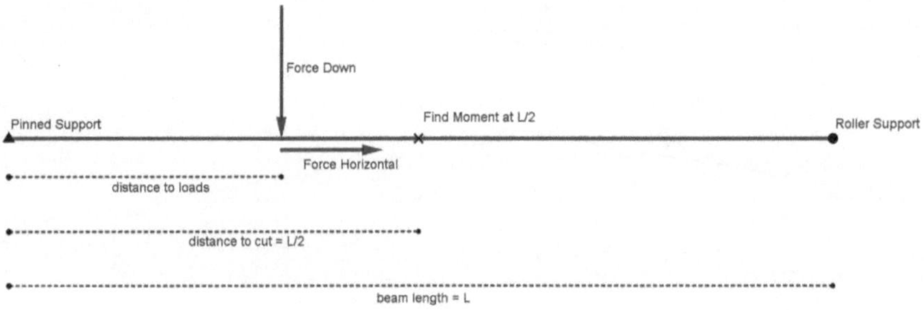

2.29
Finding internal moment at midspan of a simply supported beam

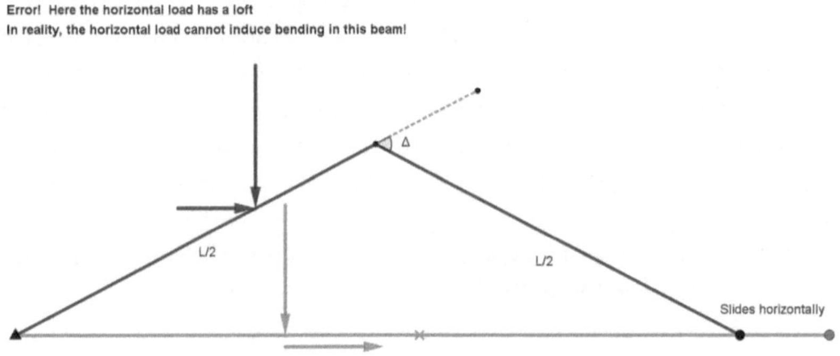

2.30
Incorrect geometry for finding the internal moment at a cut

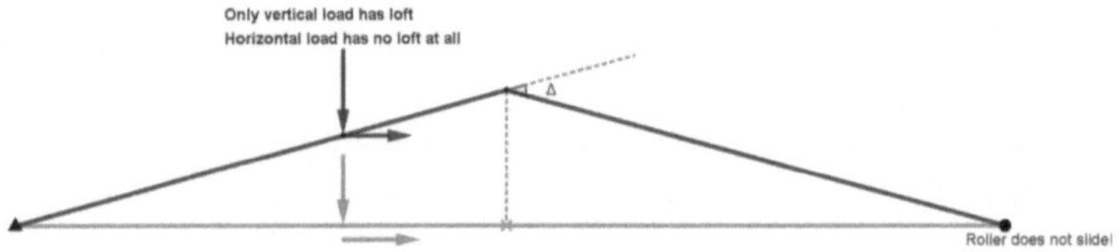

2.31
Correct geometry for finding the internal moment at a cut

The correct construction is shown in Figure 2.31.

The novice reader may think that Figure 2.31 is a violation of one of the guiding principles, as that novice reader may think that two, rather than one, elements are being deformed or stretched. But this is not a paradox. Only one element is being studied here, and the Unknown being sought is the moment at midspan of this element. Thus, this element is being deformed and stretched, each of the perturbed pieces must be longer than L/2 in Figure 2.31. But we will see in the chapter devoted to arches, that if multiple members are connected in a structure, and if we seek the moment in one element, only that element is stretched, on either side of the cut. Neighboring elements cannot stretch due to the perturbation. No paradox, no contradictions here.

Recall from Figures 2.23 and 2.24 that two beams can be connected with an internal hinge. That hinge cannot transfer bending from one member to another; that is why it is called a hinge, and it cannot transfer moment. But it does indeed transfer forces, otherwise it would be useless. It transfers two forces, much as the external "triangle" support has a vertical and horizontal load carrying capabilities. Internally, these forces have specific names. In a horizontal beam, the vertical force is known as "shear" and the horizontal force is known as "axial". These forces are important, but they rarely govern any design of a beam, because bending is far more damaging to beams than shear or axial forces are. Nevertheless, it is worth studying the shear force a bit in a simply supported beam.

The technique for establishing internal shear in a beam, at some cut a distance "*a*" from the left end of a beam whose total length is "*a* + *b*" where "*b*" is the distance from the cut to the right end, is the same as the other techniques. If seeking shear at the cut, remove the beam's ability to carry shear, perturb the cut along the direction of shear (typically vertical), and perturb it by an amount Δ. However, the perturbed shape of the beam depends on where the cut is, so the loft above the cut is shown in Equation 2.7.

$$Loft\ above\ shear\ cut = \frac{b}{a+b}\cdot\Delta \qquad \text{(Equation 2.7)}$$

And the loft below the cut is shown in Equation 2.8.

$$Loft\ below\ shear\ cut = \frac{a}{a+b}\cdot\Delta \qquad \text{(Equation 2.8)}$$

This is demonstrated in Figure 2.32. Again, there is no paradox here, as there is only one element being studied, and it is that element that will stretch. It will stretch on either side of the cut, such that the cut remains vertical.

To apply this rule to find shear at some cut, consider a beam depicted in Figure 2.33 that is 10 units of length long, with two concentrated point loads as shown. The shear at a distance "*a*" of 4 units from the left is sought.

Here, the total perturbation Δ was arbitrarily set to a large number of 0.7, but since the loads are only vertical, the answer is exact even for a large Δ. Applying Equation 1.1 of Chapter 1, and noting that Force1 induces positive work since its Loft is in the same direction as the Force and that Force2 induces negative work since its Loft is in the opposite direction of the Force, the answer is found via Equation 1.1 of Chapter 1. The geometric solution, showing the theoretical answer of 50 units of force is shown in Figure 2.34 and Equation 2.9.

$$Unknown\ Shear\cdot\Delta + Force1\cdot Loft1 - Force2\cdot Loft2 = 0 \qquad \text{(Equation 2.9)}$$

Shear will not be emphasized in this textbook, as it rarely has an effect on the final design of the structure. Much more emphasis must be placed on the internal

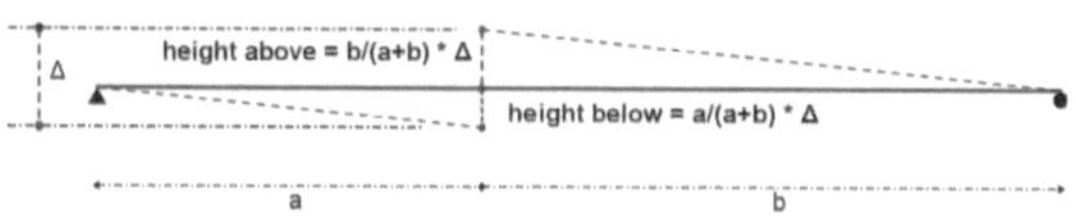

2.32
Perturbation to find internal shear at some cut in a simply supported beam

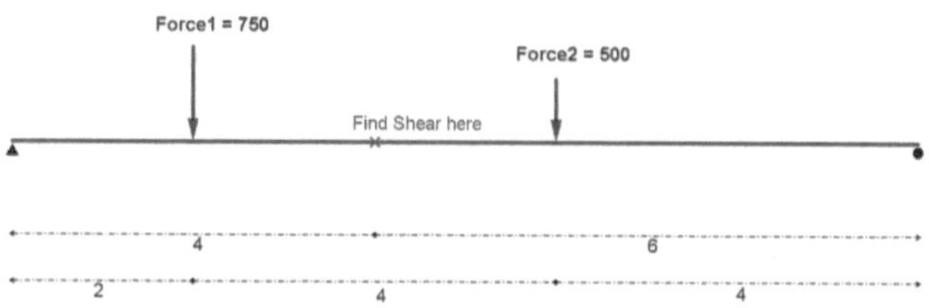

2.33
Seeking shear at some cut in a simply supported beam

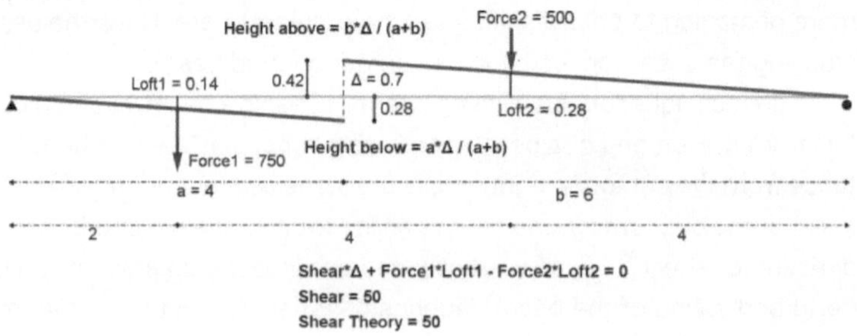

2.34
Solution for shear at some cut in a simply supported beam

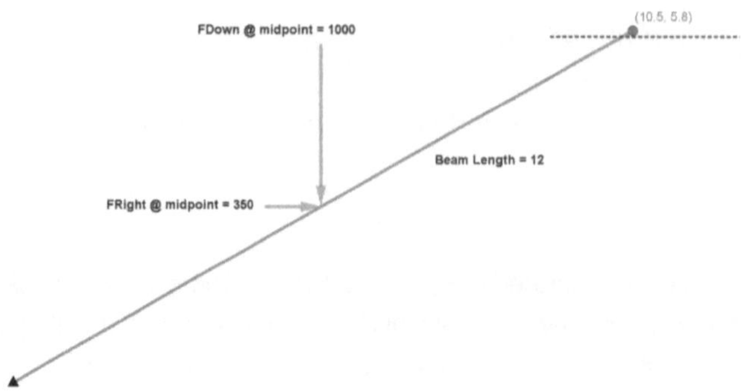

2.35
Slanted beam simply supported with two loads at midpoint

equilibrating moment at some cut in a beam. This has a profound effect on the beam's performance. As such, we will study a slightly more complicated example of a bending moment at some cut in a beam. The next example is a determinate, single beam subjected to a downward vertical load and a rightward horizontal load, both applied at the beam's midpoint. The beam is pinned at its left end, and it is vertically supported by a roller at its right end. We seek the internal bending of the beam at this midpoint, as we intuit that this will be the cut with the worst possible bending, much as if we were on a ladder, the ladder bends only a little if we are partly up the rungs, but the ladder would bend the most if we stood on a rung that was halfway up the length. The setup is shown in Figure 2.35.

Figure 2.36 shows the geometric solution for the moment at this midpoint of the beam from Figure 2.35. Notice there is a perturbation Δ which is the relative angle between the pieces on either side of the cut. In the original setup, that angle is zero. Also notice that there is only one element in this problem. We make the cut to find the Unknown moment, and we stretch each portion of that single element to ensure there is no gap between the pieces at the cut. Also notice that both loads have a vertical loft and a horizontal loft, but we only apply the vertical loft to FDown and we only apply the horizontal loft to FRight. Then Equation 1.1 of Chapter 1 immediately establishes the Unknown moment as shown in Equation 2.10.

$$\textit{Unknown Moment} \cdot \Delta - \textit{FDown} \cdot \textit{LoftVert} - \textit{FRight} \cdot \textit{LoftHoriz} = 0 \qquad \text{(Equation 2.10)}$$

As Δ gets small, the answer approaches the theoretical value of 3259 $\textit{force} \cdot \textit{length}$.

Figure 2.37 shows the moment at midspan calculation for the previous beam of Figure 2.36, but drawn as a hand sketch to show the simplicity of the method. The perturbation Δ should be large enough to draw with some accuracy, but small

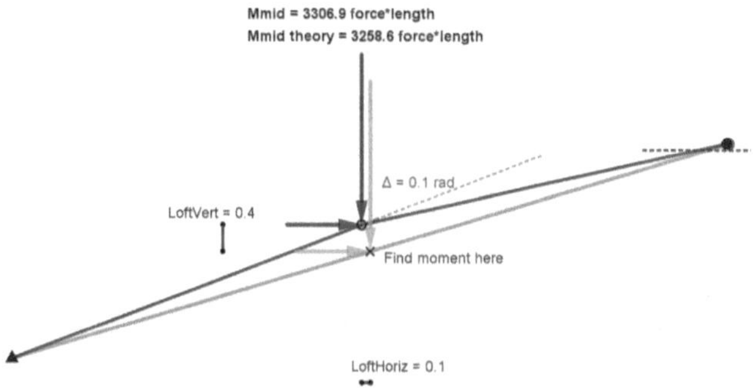

2.36
Moment at midspan for large perturbation

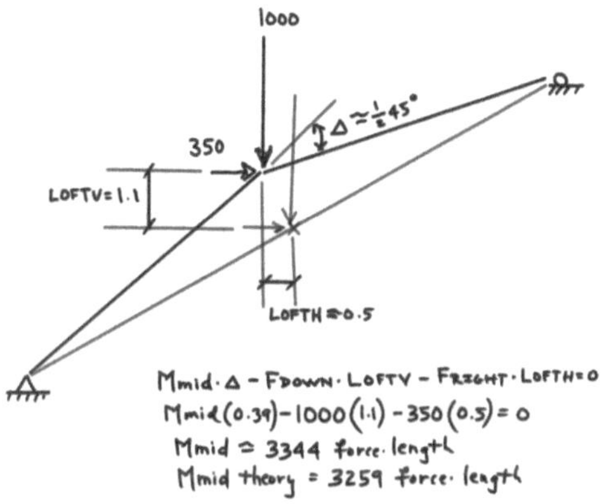

2.37
Hand sketch for moment uses ½ of 45°

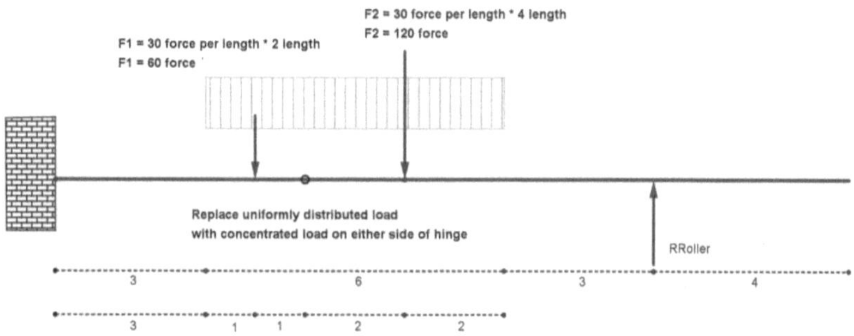

2.38 Seeking equilibrating moment at left fixed support

enough to avoid the inaccuracies of large perturbations. An angle of 22.5° is recommended for hand sketches as it is one half of 45°, and most people can draw a 45° reasonably accurately, without a protractor or rulers. The "rise" equals the "run" for a 45° slope, making it quick to draw.

Next, the propped cantilever beam with an internal hinge, originally shown in Figure 2.23 will be studied again. This time, we seek the equilibrating moment in the left fixed support. The setup is shown in Figure 2.38.

Since we are seeking the moment at the left fixed support, that is the unknown which must be removed, and a rotation Δ is imposed. All boundary conditions must be maintained, but notice the element with the unknown, namely the beam spanning from the left fixed support to the internal hinge is stretched, such that the hinge drops down vertically without any horizontal movement. This means that the lofts for the loads broken up proportionally to either side of the kink at the hinge are directly below the equivalent concentrated forces to the left and to the right of the kink at the hinge. Equation 1.1 establishes the Unknown Moment in the Left Wall. The Δ and the two Lofts are clearly shown in Figure 2.39 and Equation 2.11.

$$MLeftWall \cdot \Delta + F1 \cdot Loft1 + F2 \cdot Loft2 = 0 \qquad \text{(Equation 2.11)}$$

One detail that is not influential but is worth noting in Figure 2.39 is that the element being cracked is the beam from the wall to the hinge. As such, that element is stretched. But the remaining portion of the beam is not stretched. This is gently noted by the length of the element being shown as 10 units of length, and by a very slight shift of the load on the perturbed beam, which must be 2 units of length to the right of the hinge. As the perturbation Δ gets small, such shifts become zero.

For a small Δ, as shown in Figure 2.40, the answer approaches the exact theoretical value of −668 force length.

The case of a concentrated externally applied moment has not yet been studied in this chapter. It may seem a bit academic to look at a situation such as the one shown in Figure 2.41, but there is a very good reason why we should study it. We will use this concept in étude 9.3. For now, in Figure 2.41, we see a simply supported beam, of span *L*, with no loads at all applied between the supports. However, a concentrated, externally applied moment MLoad is applied at the left end. We seek the left vertical reaction.

The reason why this problem is worth studying will become apparent only when the study of indeterminate beams has begun. Then, the reader will see that one drawing will be needed to find the equilibrating moment in the so-called "propped cantilever" beam shown in Figure 2.42. The indeterminate propped cantilever is subjected to external loads, and in the chapter on Indeterminate Structures,

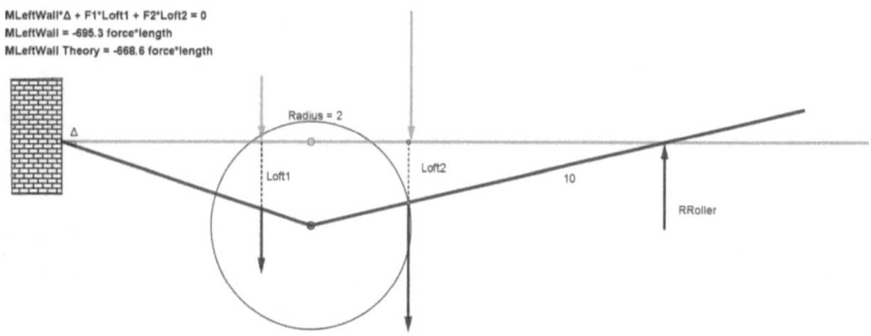

2.39
Establishing equilibrating moment at left fixed with large Δ

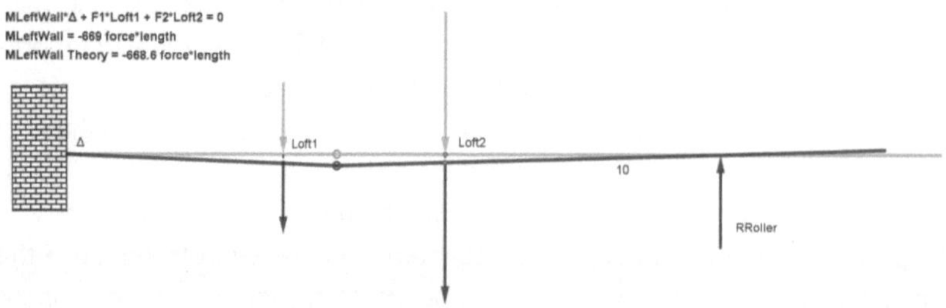

2.40
Establishing equilibrating moment at left fixed with small Δ

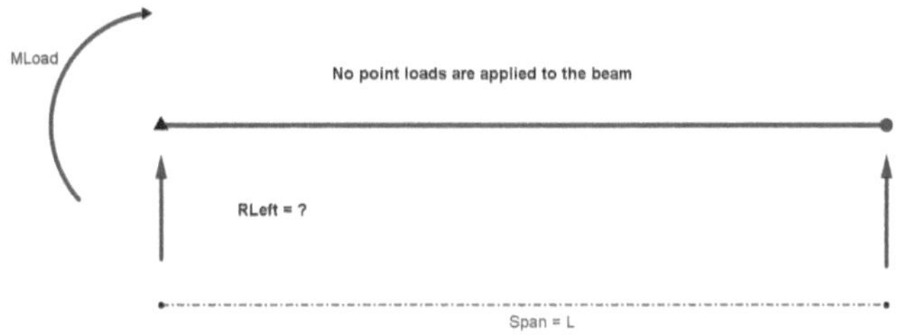

2.41
Simply supported beam with concentrated moment as the externally applied load

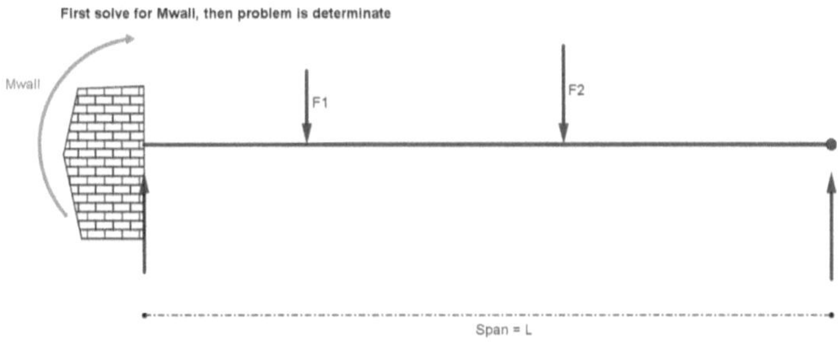

2.42
Indeterminate propped cantilever

the reader will learn how to solve for the equilibrating moment in the fixed wall. Once that moment is found, the problem is unlocked and becomes determinate. Then, a second drawing will establish any equilibrating unknown in the now determinate beam.

Thus, we return to a determinate problem similar to the one shown in Figure 2.42, where an externally applied moment is applied. Suppose we sought the left equilibrating vertical reaction for the structure shown in Figure 2.43. Since we use the Modern Müller-Breslau Method to establish a reaction, we do not stretch any elements, rather the perturbation forces elements to move in a circular path. A coarse solution is shown in Figure 2.43. Notice that the loft associated with the applied external moment is a slope, not a movement.

And the solution approaches the theoretical value of −749.5 units of force, as Δ is reduced. This is shown in Figure 2.44.

The most common statically determinate beams have been described in this chapter. The setup of solving for external reactions and internal equilibrators, be they forces or moments, has been emphasized throughout this chapter. The reader is once again strongly encouraged to sketch out by hand the various solution

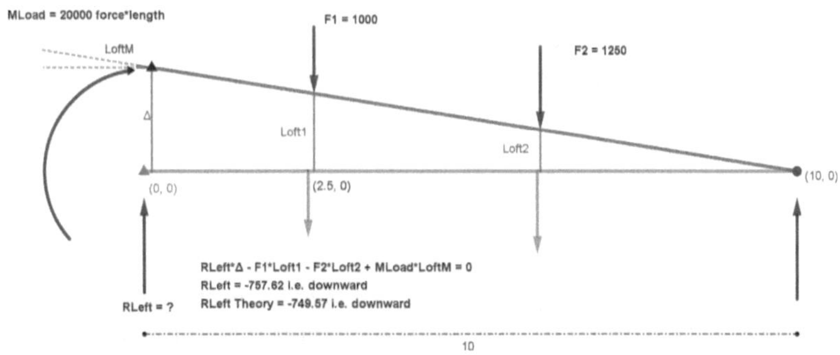

2.43
Determinate beam subjected to external moment and two external loads

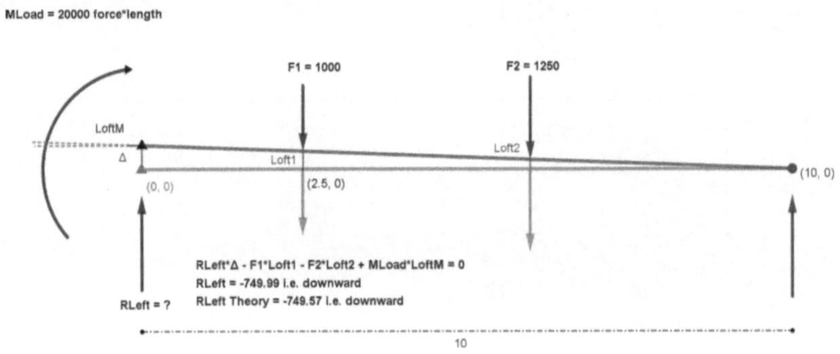

2.44
Solution for RLeft

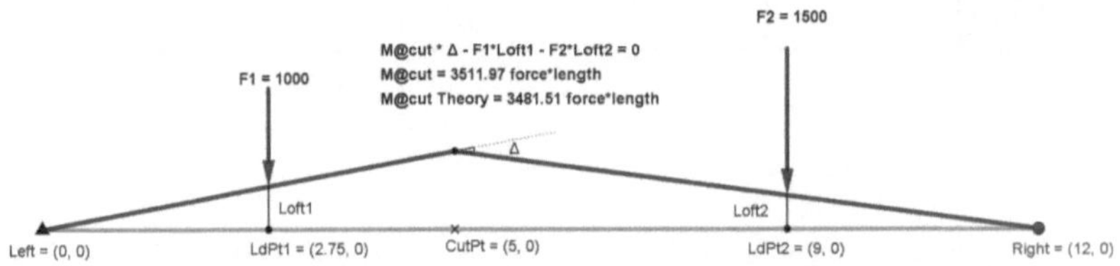

2.45
Setting up an automatic graph of bending at any cut

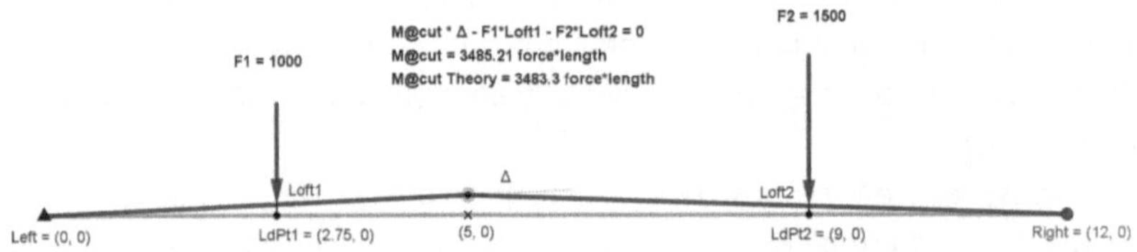

2.46
Magnitude of the moment at one cut is plotted

geometries in this chapter, with no emphasis placed on quantitative answers. Rather, it is important to get a sense of how to construct the perturbed shapes. In the next chapter, many details will be provided on how to draw these perturbed shapes in a parametric computer program.

We close this chapter with a surprise! The bending moments can be charted or graphed for any cut in the beam. This is known as a "bending moment diagram" and experienced readers will be very familiar with bending moment diagrams. The surprise is that they can be drawn with very little extra effort for simple problems. For more complicated problems, a few parametric drawing puzzles need to be solved. As an example, consider the simply supported beam with two concentrated point loads at fixed and known positions. We initially establish the moment at some cut, here the cut is at $x = 5$ as shown in Figure 2.45.

With a small perturbation Δ, the answer approaches the theoretical value. This is shown in Figure 2.46. Notice in Figures 2.45 and 2.46 how the magnitude of the bending moment at the cut would need its *own scale* if it were graphically drawn on the beam.

Now here is the surprise! With a little bit of programming, the cut can be moved anywhere along the length of the beam and the trace of the moment versus any cut x can be displayed. The trace remains on the screen as the cut being investigated changes. This is the bending moment diagram, drawn to some comfortable scale. A bending moment diagram is shown in Figure 2.47. Notice it is a funicular, i.e. the

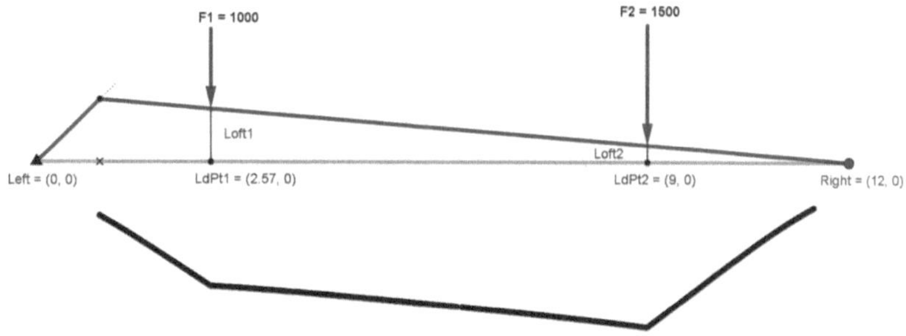

2.47
Bending moment is plotted along the beam length

shape that a hanging chain would take under any loads. Here, the chain would be straight between the two-point loads, as we neglect the self-weight of the theoretical chain. The chain's sag would be more pronounced under the heavier load. All of this is confirmed by our intuition of such hanging cables found in everyday life.

CHAPTER 2 ÉTUDES

The following studies or études should be sketched by hand.

Étude 2.1) Draw a beam of $L = 12$ between a pin support at the left and a roller support at the right. There is a 4 unit length overhang to the right of the roller, creating a total beam length of 16. We seek the left vertical reaction, draw the perturbed shape for a Δ of 3 using the MMB, i.e. no stretching of the element. Note that for a Δ of 6, the rotation would be 45°, so use that as a guide and then take ½ of 45°. Put a downward load of 1,000 at 6 units of length from the left pin and put a downward load of 500 at the free right tip. Calculate quantitatively the vertical left reaction. A hand sketched solution is shown in Figure 2.48.

Étude 2.2) Draw a beam with no overhangs, which is roller vertically supported at the left and pinned at the right. But the beam is slanted such that the left is at (0,6) and the right is at (10,0) We seek the left vertical reaction, draw the perturbed shape for a Δ of 2.0 using the MMB, i.e. no stretching of the element. Put a single leftwards horizontal load of 1,000 at the midpoint of the beam. Find the left vertical reaction for this very large perturbation. A hand sketched solution is shown in Figure 2.49.

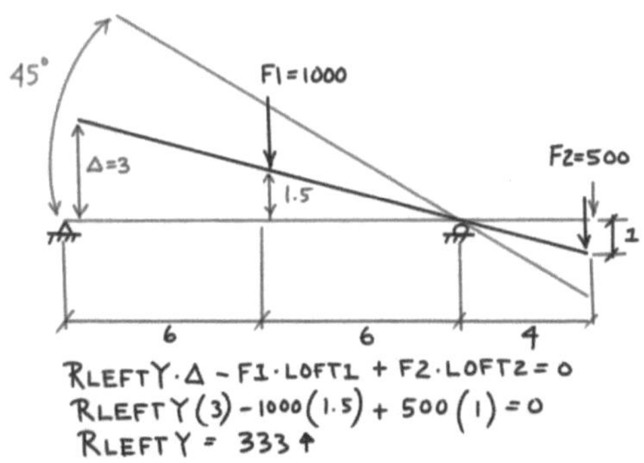

2.48
Étude 2.1 solution

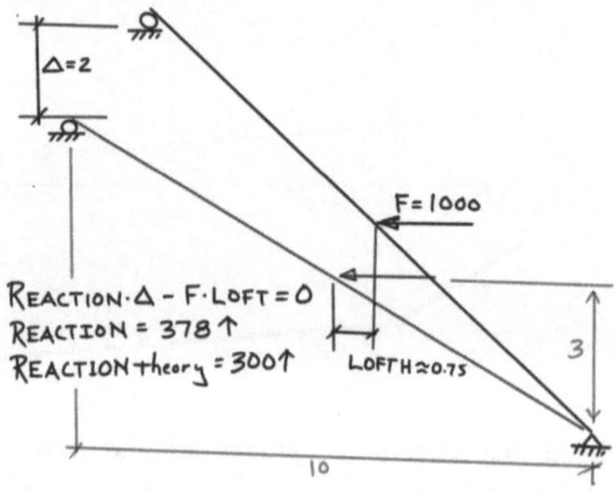

2.49
Étude 2.2 solution

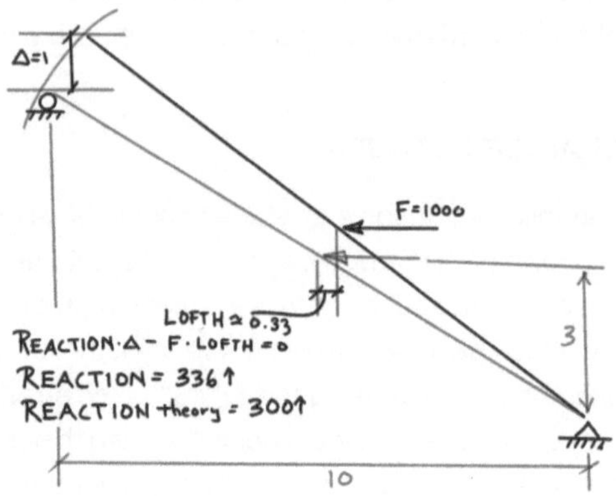

2.50
Étude 2.3 solution

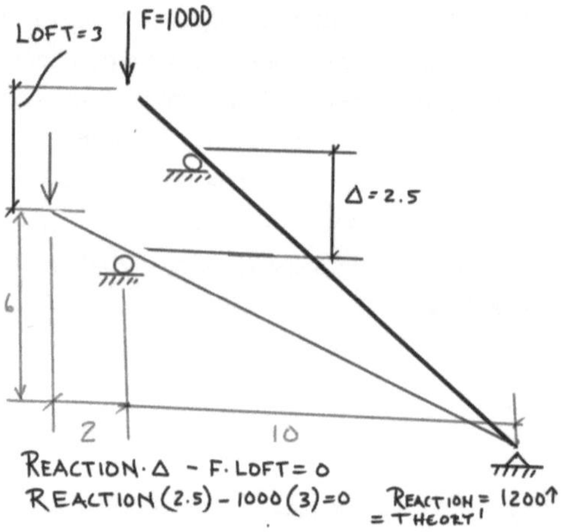

2.51
Étude 2.4 solution

Étude 2.3) Repeat Étude 2.2 with a Δ of 1. A hand sketched solution is shown in Figure 2.50.

Étude 2.4) Draw a beam with one overhang. A roller at (2,5) is vertically supporting the beam at the left and pinned at the right at (12,0). But the beam is slanted such that the free overhanging tip at the left is at (0,6). We seek the left vertical reaction, draw the perturbed shape for a Δ of 2.5 using the MMB, i.e. no stretching of the element. Put a single downwards vertical load of 1,000 at the free left

2.52
Étude 2.5 solution

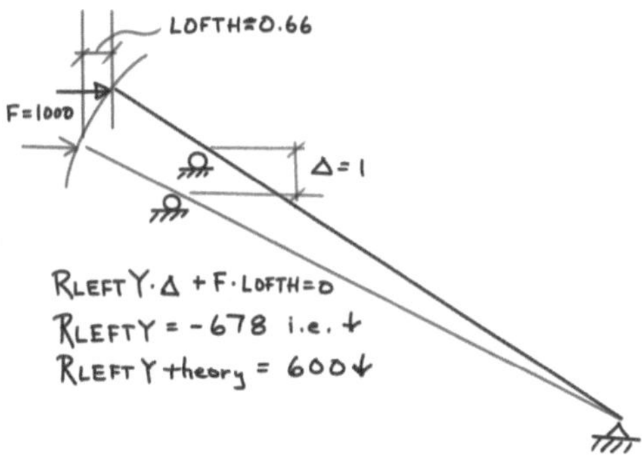

tip. Find the left vertical reaction for this very large perturbation. A hand sketched solution is shown in Figure 2.51.

Étude 2.5) Repeat Étude 2.4, but replace the vertical load with a rightward load at the free left tip of 1000. Use a Δ = 1. Find the approximate left vertical reaction and note whether it is upwards or downwards. Verify the direction with your intuition. A hand sketched solution is shown in Figure 2.52.

Étude 2.6) A beam is pinned at the left end and roller supported vertically at the right end. The total length of the beam is 12 units of length. There is a vertical appendage which begins at 4 units to the right of the left support. That appendage is 3 units in length tall. At the free tip of the appendage is a horizontal load to the right with a magnitude of 1,000 units of force. Find the right equilibrating vertical reaction. Draw Δ = 2 to get a reasonably good answer. Be sure to specify the direction of the equilibrating force, i.e. up or down. A hand sketched solution is shown in Figure 2.53.

Étude 2.7) Repeat Étude 2.6, but this time find the left vertical equilibrating reaction. Be sure to specify the direction of the equilibrating force, i.e. up or down. Does your answer make intuitive sense? A hand sketched solution is shown in Figure 2.54.

Étude 2.8) Repeat Étude 2.7, but this time find the left horizontal equilibrating reaction. Be sure to specify the direction of the equilibrating force, i.e. left or right. Does your answer make intuitive sense?

2.53
Étude 2.6 solution

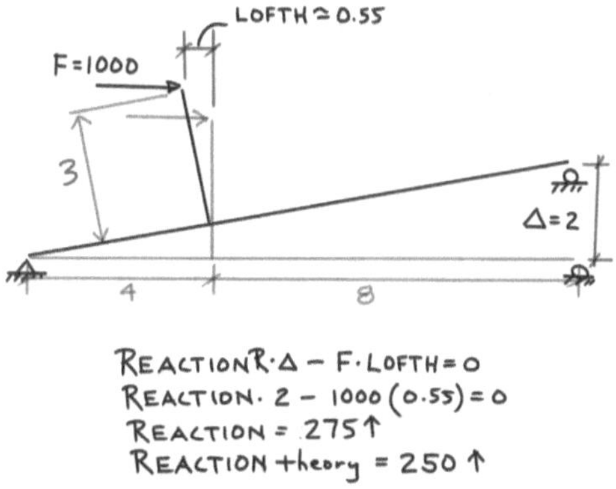

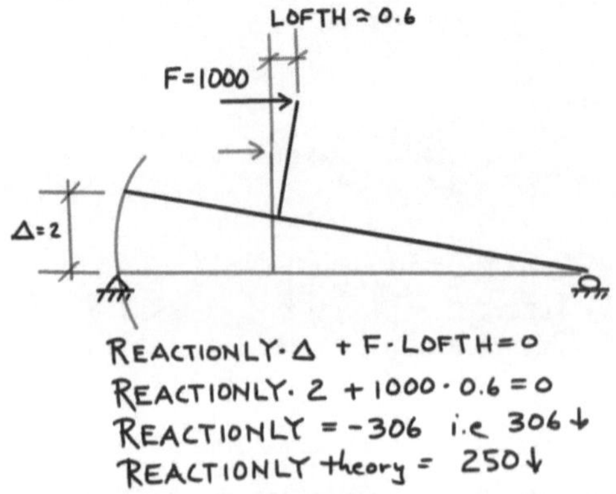

2.54
Étude 2.7 solution

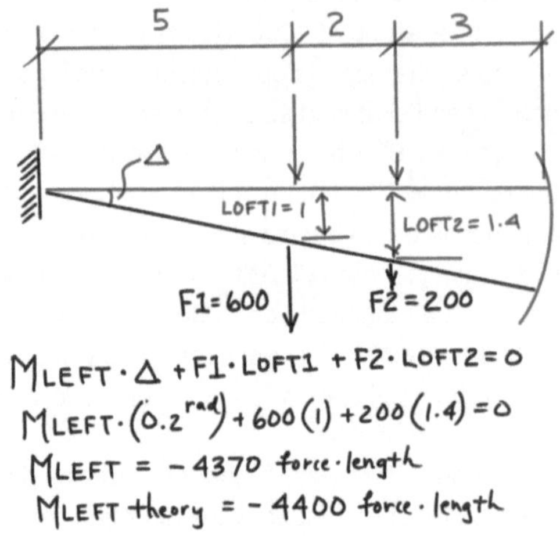

2.55
Étude 2.9 solution

Answer: No graphics needed. The equilibrating horizontal pin force must be 1,000 going to the left.

Étude 2.9) A cantilever beam has a total length of 10 units of length. It is fixed at its left end. There are two downward point loads. The first load is 600 units of force applied at 5 units of length from the left fixed end. The second load is 200 units of force applied at 7 units of length from the left fixed end. Find the equilibrating moment in the wall if the loft of the 600 unit load is 1. A hand sketched solution is shown in Figure 2.55.

Étude 2.10) Repeat Étude 2.9 but this time make the loft under the 200 unit load equal to 1. A hand sketched solution is shown in Figure 2.56.

Étude 2.11) A propped cantilever is fixed at the left end and roller supported vertically at the right end. There is an internal hinge in the beam at 5 units of length from the left wall, with 7 units of length right of the hinge to the right roller, for a beam length of 12 units. Find the right reaction if the entire beam is loaded with 100 force/length of uniformly applied downward load. Try a perturbation of $\Delta = 3$ to get a visibly large lofted structure. A hand sketched solution is shown in Figure 2.57.

2.56
Étude 2.10 solution

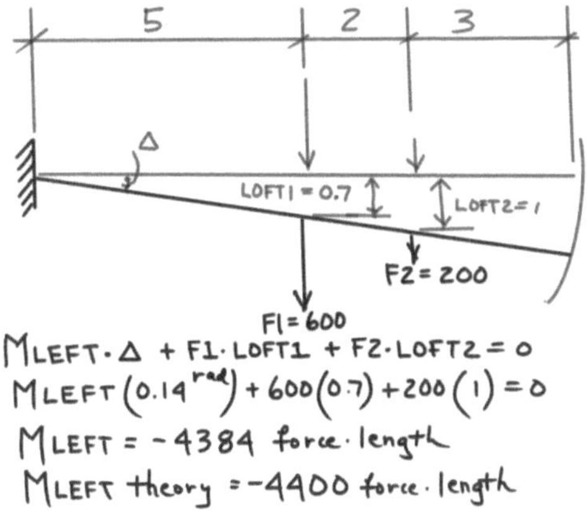

2.57
Étude 2.11 solution

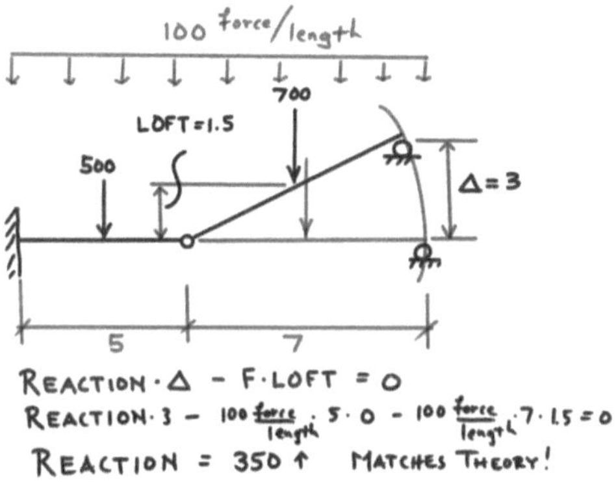

2.58
Étude 2.12 solution

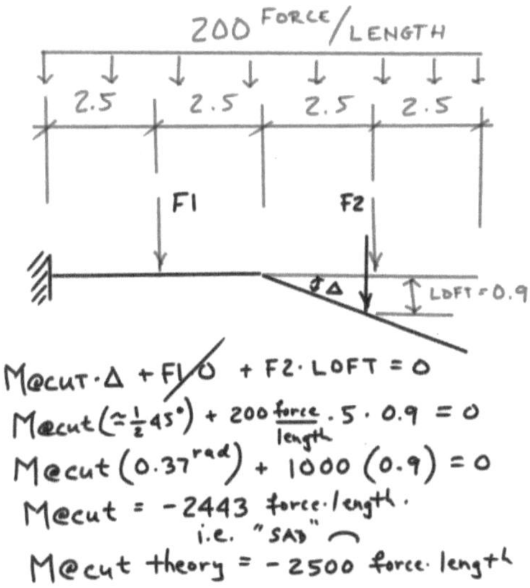

Étude 2.12) A cantilever beam of length 10 is fixed at its left end, and free at its right end. The entire beam is loaded by a downward, uniformly distributed load of 200 force/length. Find the moment at a cut at the mid-span of this beam. Include

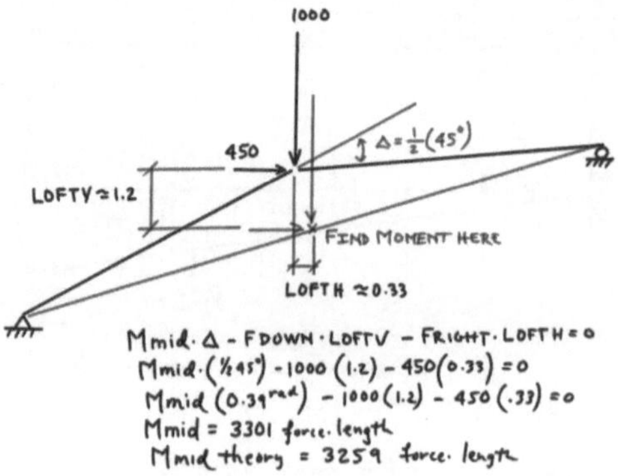

2.59
Étude 2.13 solution

the sign of the moment in your answer, i.e. is it "happy" (concave up or positive) or "sad" (concave down or negative). A hand sketched solution is shown in Figure 2.58.

Étude 2.13) A slanted beam is 12 units of length, with its left end pinned at coordinates (0, 0) and with its right end vertically supported by a roller at coordinates (11.5, 3.3). Find the moment in a cut at the mid-span of this slanted beam if the beam is subjected to two loads, one downward load of 1,000 units of force and one rightward load of 450 units of force, both of which are applied at the mid-span of the beam. Since there is only one element, be sure to keep the left and the right ends at their original points, i.e. stretch that single cracked beam. A hand sketched solution is shown in Figure 2.59.

Étude 2.14) A horizontal beam is 14 units of length, with its left end pinned at coordinates (0, 0) and with its right end vertically supported by a roller at coordinates (10,0). There is a 4 unit overhang to the right of the roller. The entire beam is loaded with a uniformly distributed load of 150 force/length. Also, there is a concentrated point load of 1,000 units of force at the free tip. Make the depth (or the height) of the beam 1 unit of length. We will find the moment in a cut located at 6 units of distance from the left pinned support. Initially, simply sketch the setup. A solution is shown in Figure 2.60.

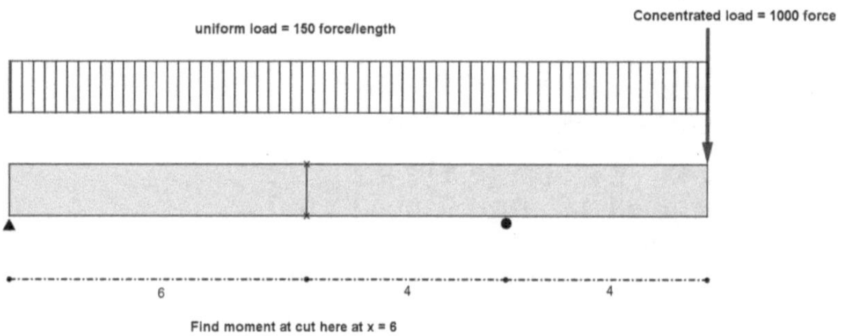

2.60
Étude 2.14 setup

2.61
Étude 2.15 solution

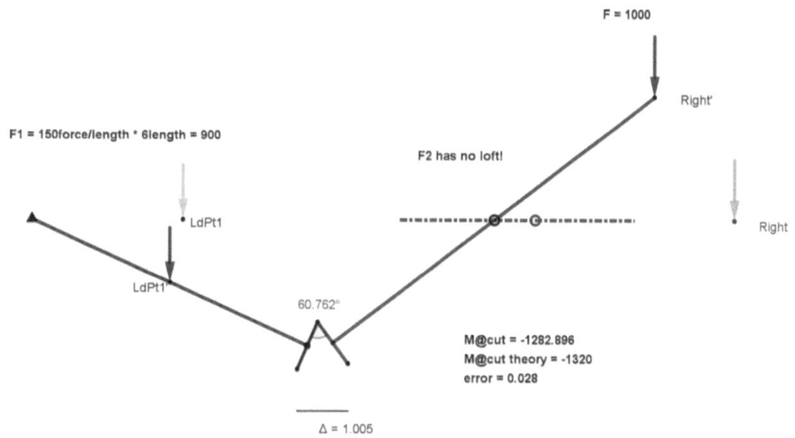

Étude 2.15) For the beam described in étude 2.14, solve for the moment in a cut located at 6 units of distance from the left pinned support. Be sure to use a depth of beam equal to 1 unit of length and make sure the "crack" or hinge is at the top of the beam. Open up the beam 1 unit of width at the bottom of the beam. Allow the right roller to slide. A solution is shown in Figure 2.61.

3 GeoGebra for Beams

To introduce the tool GeoGebra which was used to generate the geometric solutions shown up till now, the perturbed beam of Figure 2.15 in Chapter 2 will be studied in detail. It contains enough complexity, that if mastered, the reader will have a solid grasp of the parametric drawing tool GeoGebra.

All of the examples use GeoGebra Classic 5, which combines robustness with ease of use. The program should be downloaded for free at App Downloads – GeoGebra (https://www.geogebra.org/download). Windows users will have no issues installing the software, whereas Mac users should follow the advice on GeoGebra's website:

- Using Finder on your Mac, locate the downloaded GeoGebra app you want to open.
- Control-click the app icon, then choose Open from the shortcut menu.
- Click Open.
- The app is saved as an exception to your security settings, and you can open it in the future by double clicking it, just as you can with any authorized app.

Figure 2.15 of Chapter 2 is re-created here as Figure 3.1, with a bit more detail shown.

Before drawing anything in GeoGebra, it is important to become familiar with the main functional windows of the program. Figure 3.2 shows a typical representation of a new setup. In Figure 3.2, the three main parts are the Algebra Window

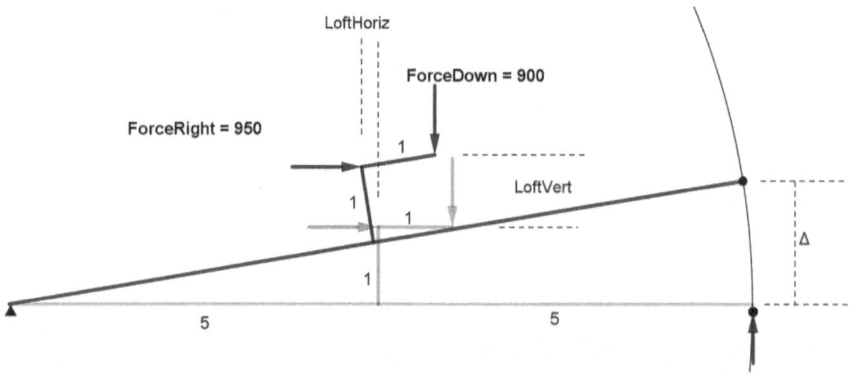

3.1
Some details of an indirectly loaded beam

DOI: 10.4324/9781003180913-3

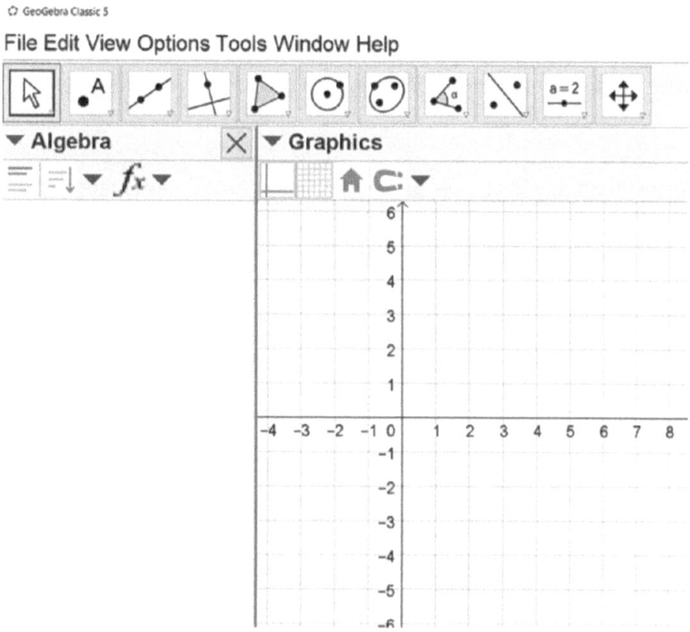

3.2
Initial screen of GeoGebra Classic 5

which displays the actual information being programmed, the Graphics Window, which is a two-dimensional representation of the information being programmed (3D graphics are in a separate window), and the Input Bar where commands are actually typed in. Other features exist in GeoGebra, and of particular note is the CAS Window, which stands for Computer Algebra System, an extremely powerful symbolic toolkit, which will not be studied in this textbook.

Several quick hints will make the reader's experience much more pleasant:

- Options >> Labeling >> No New Objects is very handy, avoiding the cumbersome labels appearing on each new item.
- Closing the Algebra Window, then re-opening it immediately, allows for quick manipulation of Object Properties in the Graphics Window, for example easily turning on and off the grid.
- Sorting the items in the Algebra Window, by Object Type, rather than by Order of Appearance is very useful.
- Points must be created using an initial capital letter.
- Overall font size can be easily increased for ease of viewing, Options >> Font Size.

These initial hints are sufficient to get novice users over the initial learning curve hump, which is very gentle indeed.

To create the initial beam of Figure 3.1 in GeoGebra, first recognize that of course there are many methods to draw any entity, but the ease of the following technique might make it the preferred method.

All the icons along the top bar have sub-menus under the initial menu. For example, under the third top icon, the default is the Line tool, but under that is the Segment tool. Use a Segment to establish an image of the beam connecting the

Left and Right points. Right click on the Segment to change thickness, line style, opacity, and color.

Here are a few more general hints that may make the work proceed quickly and efficiently:

- Significant items such as lengths of spans, magnitudes of loads, lengths of cantilevers, etc. should all be pre-programmed using realistic names in the Input Bar and they will appear in the Algebra Window. For example, Force1 = 1,000, CantileverRight = 6 are clearly named, and are case-sensitive, meaning capitalization matters. Do not use spaces within a given name. Never use vague names to define something significant. But don't bother renaming all the insignificant points, segments, and lines, most of these will eventually be hidden.
- Hide, do not delete! Deleting an entity also deletes all entities that depend on the original item. Simply hide the many superfluous items that will arise, such as construction lines, to make the drawing more clean and tidy.
- The midpoint button under the Point Menu is extremely easy to use and very helpful.
- Show the axes originally, then draw parallel or perpendicular lines as needed, but quickly hide the axes and hide the grid for a clean drawing.
- Hide absolutely everything that is not essential, and there will be many such items.
- When searching for something like the name of a point at the end of a segment, right click on the items and look at the Basic Definition, which always describes how the item was constructed.
- Circles which intersect lines are a terrific way of locating points in 2D space. This is particularly convenient as the radius of the circle needs to be predetermined with pre-programmed parameters. For example, the size of a load vector could be scaled by a circle whose radius is Force1/ForceScale where ForceScale is an arbitrary number that comfortably fits the vector describing the load applied to a beam.

Another very useful way of creating entities is to refer to the first entity and simply add a parameter to it. For example, start to recreate Figure 3.1 by typing the definition of the left support in the Input Bar, for example Left = (0,0). There is no need to input the third dimension, only the *X* and *Y* Cartesian Coordinates are needed, using parentheses around two comma separated numbers. Then in the Input Bar, for example type Length = 10. Notice that there are no units in GeoGebra, neither Imperial nor Metric. The user is responsible for ensuring consistency of units. Then in the Input Bar, typing in Right = Left + (Length, 0) is a quick and parametric way of establishing the Right point, with the parentheses enclosing the x coordinate and the y coordinate, respectively. Changing the variable Length in the Algebra Window will automatically update the position of the Right point.

Using these helpful hints as a guide to constructing the perturbed beam of Figure 3.1, it makes sense to draw a Segment between the pre-defined Left and Right points, with the definition of the Right point as being dependent on the Left point

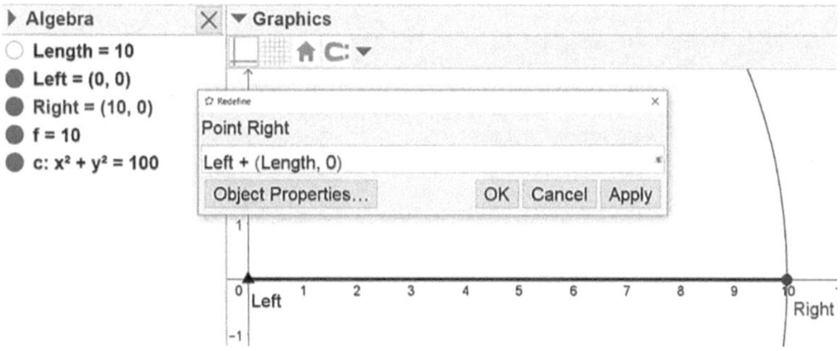

3.3
Start of drawing of Figure 3.1

as well as the parameter Length. Don't worry about the "triangle" and the "roller" being slightly below the beam, that can come later. Then draw Circle: Center and Radius, not Circle: With Center Through Point, to draw a circle which is centered on the left support and has a Radius of the previously defined Length value. Figure 3.3 shows the development of these ideas. Note in Figure 3.3 that double clicking on an item such as Right in the Algebra Window opens up a small pop-up window that displays the definition and allows for redefining the parameter.

Notice in Figure 3.3 that the label for the horizontal segment representing the beam is hidden. Also, the label for the circle is hidden, but the labels and icons of the Left and Right points are intentionally displayed.

Snapping a point to the circle is extremely straightforward, simply choose the Point icon and touch anywhere on the drawn circle. Note this is much more difficult to do in many other programs such as Grasshopper, here in GeoGebra the snap is "a snap!" Switch to cursor mode and note that the Point is now glued to the circle, try moving it around. Then a second beam can be drawn, and the perturbation Δ can be calculated at least two ways. One way is to draw a vertical segment between the perturbed end and the original horizontal line. The length of the segment need not be calculated as Length(item) in the Input Bar, where "item" is the name of the segment, although that does work. It is much easier to simply refer to "item" which is the name of the segment, and GeoGebra interprets that as the scalar length of the segment. This is a visual affirmation of the actual value of Δ, but it does require the construction of that segment. A more direct manner is to simply calculate the difference in the y value of the perturbed point and the y value of the unperturbed point. All of this is shown in Figure 3.4. Note that in Figure 3.4, Point A is at some random large perturbation. The reader should recreate Figure 3.4 and try to get the same Δ as shown, by wiggling Point A along that arc. Clearly, for different positions of Point A, differing Δ values exist.

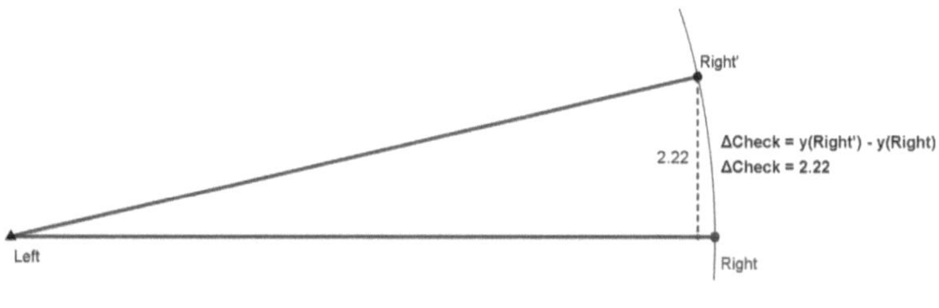

3.4
Two ways of establishing Δ

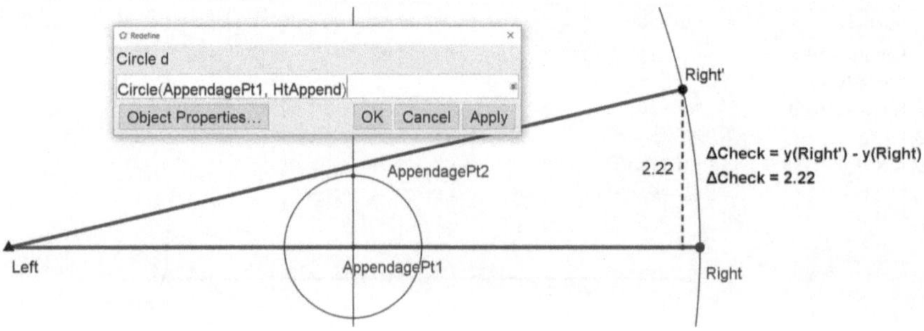

3.5
Two ways of establishing the first leg of the appendage

There are at least two efficient ways of drawing the "appendage" of Figure 3.1. Notice that the appendage originates at the halfway point along the span of the beam, i.e. 5 units of length from the Left end. A quick way of establishing the midpoint between any two points is to use the Midpoint or Center icon, under the Point drop down menu. Choosing the Midpoint icon, then clicking on the previously defined Left and Right points quickly establishes the mid-way point which becomes the origin of the appendage. Of course, that point could have been explicitly stated as AppendagePt1 = (XAppend, 0) where XAppend is a previously defined value describing the origin of the appendage. Then there are two straightforward ways of finding the top of the first appendage. Either enter AppendagePt2 = AppendagePt1 + (HtAppend, 0) where HtAppend is a previously defined height of the first appendage (here that value is 1 unit of length), or draw a circle of radius 1, and find the 12 o'clock intersect with a vertical line through AppendagePt1. Both techniques are displayed in Figure 3.5.

Establishing the second leg of the appendage can follow the same technique as the drawing of the first appendage. That is, one could locate CCC = CC + (HtAppend, 0) since the second leg of the appendage has the same length as the first leg. Or, another circle could be drawn to find point CCC. These methods are both shown in Figure 3.6. The reader should be certain of the following fact. Either of the methods of locating Point CCC in Figure 3.6 is vastly superior to the incorrectly used CCC = (6,1). Why is simply "hardwiring" the Cartesian Coordinates of CCC to be highly avoided? For at least two reasons. The first is that forcing CCC to always be located at (6,1) means that the program is no longer truly parametric, it cannot be easily manipulated and reconfigured. The second reason is that it is much more difficult to "debug" or to find errors in a program that has such fixed values buried

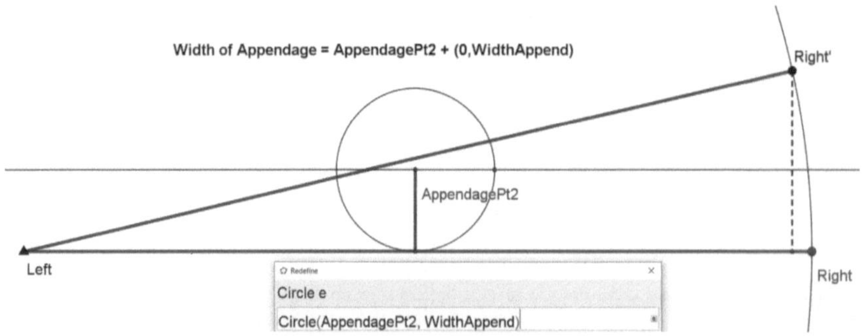

3.6
Two ways of establishing the second leg of the appendage

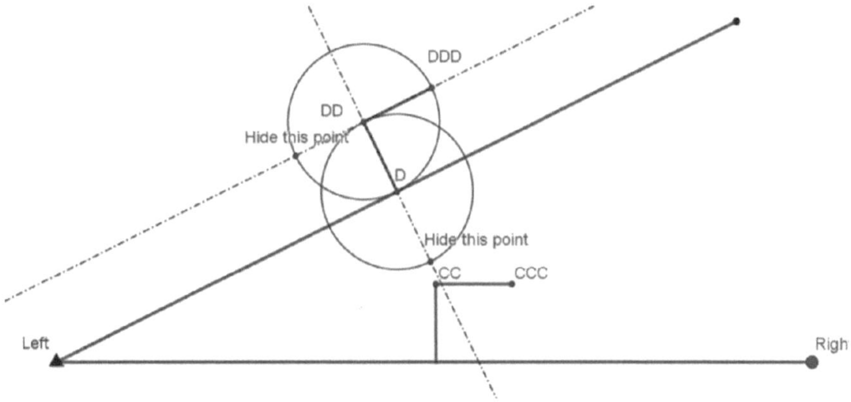

3.7
Intersect a line and a circle to locate a point

somewhere. As your programs get more complicated, it will become nearly impossible to trace down such "hardwired" entities, so please avoid them and make everything as parametric as possible.

On the perturbed beam, the exact same technique can be used to draw the new appendages in their new position. This is shown in Figure 3.7. Note that anything connected to the perturbed beam, be it the appendage or the loads, will swing along the circular path established at the onset of the problem. Check to see that when the new Right end moves along the arc, everything moves along the beam. Also note that the preferred method of simply using a circle and an intersecting line to find point DD is really much easier than the other method of finding DD = D + (#, #) since the *X* and *Y* perturbations are not immediately obvious! Always choose the simpler, nearly foolproof path of using a circle and an intersecting line when in doubt. Obviously, note there are two intersecting points, but simply hide the 6 o'clock point, do not delete it as deleting it will delete the 12 o'clock point as well. This is shown in Figure 3.7. While it is not strictly necessary to alter line weight, as was done in Figure 3.7, doing so is very helpful when understanding more and more complicated drawings. Line weight and other options can be adjusted with a right click and then Object Properties. Typically, heavier line weight is used for primary elements such as beams. Lighter line weight or dashes are used to show where points lie in 2D space, these are called "construction lines" as they help the user construct beautiful drawings. The reader should take pride in creating clean, easy to read and aesthetically pleasing digital as well as hand sketched, anthropic drawings.

The problem could be completed now, as the drawing of the load vectors is not needed for the solution. The magnitudes of the forces are needed, these should be typed into the Input Line and then clearly displayed in the Algebra Window. Having the magnitude of the vertical force, and then the vertical lofts, or the vertical distance between Point DDD and Point CCC, for example, would be enough information to calculate the negative work performed by the vertical force on CCC. But it is fun and helpful to draw the forces as vector arrows. The vertical force vector on CCC and then on DDD will be drawn as follows. First, establish the magnitude of the downward force, in Figure 3.1 this is 900 units of force, perhaps call it ForceDown. Again, note that GeoGebra does not use built-in units, neither pounds nor kilonewtons are built in; the user must keep track of the units. Then,

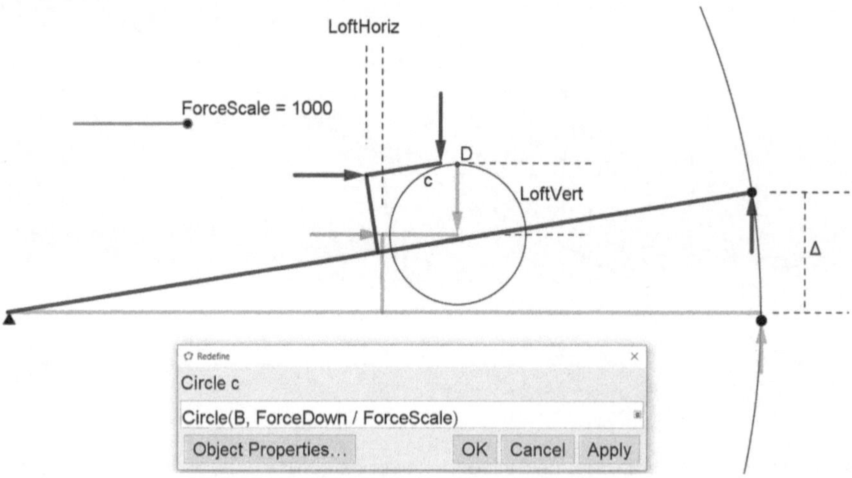

3.8
Drawing the first force vector to some comfortable scale

establish a ForceScale, perhaps on a slider for ease of adjustment. Sliders are found in the second to the right icon button along the top. A handy set of limits on ForceScale is a lower limit of 1 and an upper limit of ForceDown. Then, a circle is drawn around Point CCC for example, and the radius of that circle is ForceDown/ForceScale, where ForceScale is slid to some comfortable value. The intersects at 12 o'clock and 6 o'clock are established, and the 6 o'clock point is hidden. Then a Vector, found under the third from the left icon menu bottom along the top, is drawn. All this is shown in Figure 3.8.

Once again, notice how much easier it is to locate points using the circle and line intersection. Establishing the orientation of the perturbed downward force will reinforce that idea one more time. Recall that the downward force is indeed downward, even though it is tempting to draw it perpendicular to the perturbed beam. In all of the work in this textbook, downward loads remain downward throughout a perturbation, horizontal loads remain horizontal throughout a perturbation. Figure 3.9 shows the construction of that second vector, and the opacity of the first downward vector is lightened to show that it is no longer there.

Finally, the vertical loft of the ForceDown is quickly found in one of two ways, as described previously. Either via a segment that is vertical between horizontal lines passing through CCC and DDD, or by calculating the difference between the

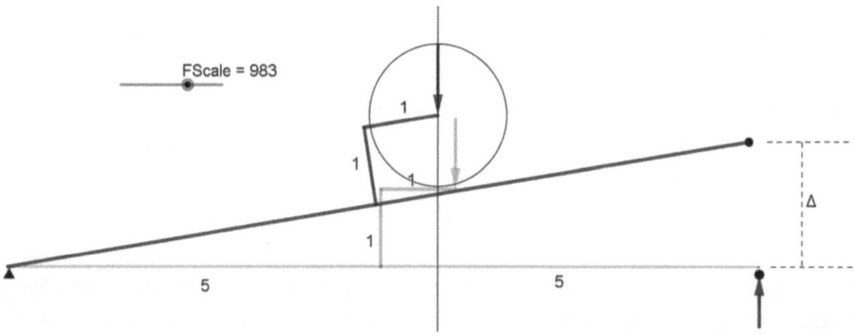

3.9
Drawing the second force vector to some comfortable scale

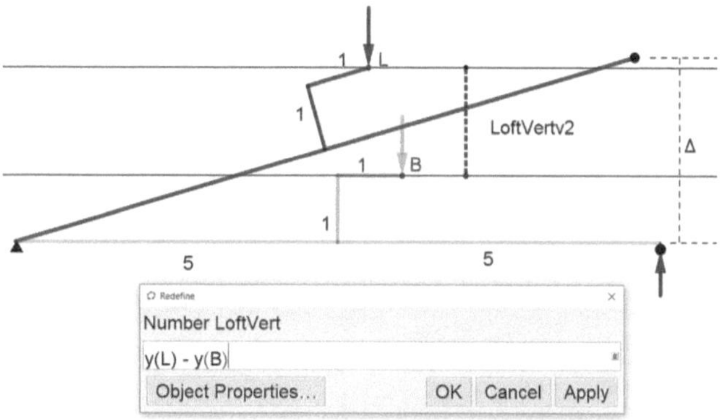

3.10
Drawing the loft and using dynamic text

two points identifying the downward load, namely via *y*(DDD) – *y*(CCC). This is shown in Figure 3.10.

These steps are repeated for the horizontal load, which is always drawn to the exact same scale as the vertical load. This is the essence of a consistent drawing scale, all entities are unified by a single factor. Never draw forces to different scales unless the loads are different by orders of magnitude. In that case of wildly different load values, be sure to explicitly state the scale factor being used to display each particular load. Figure 3.11 shows the second load drawn to the same scale as the first load. Also, Figure 3.11 demonstrates how quickly a drawing can get cluttered, proving the need to hide everything except the essential items that convey the necessary message.

Having the vertical load and the vertical loft, as well as the horizontal load and the horizontal loft, as well as the perturbation Δ, which was established in Figure 3.4, the unknown vertical reaction can now be found. The signs of the work in Equation 1.1 of Chapter 1 do indeed matter.

$$Unknown \cdot \Delta + \sum Force_i \cdot loft_i = 0 \qquad \text{(Equation 3.1)}$$

$$RightReaction \cdot \Delta - ForceDown \cdot LoftVert - ForceRight \cdot LoftHoriz = 0 \qquad \text{(Equation 3.2)}$$

$$RightReaction = \frac{ForceDown \cdot LoftVert + ForceRight \cdot LoftHoriz}{\Delta} \qquad \text{(Equation 3.3)}$$

As the perturbation gets smaller and smaller, the solution will approach the theoretical answer of 635 units of force upwards. Notice that we assumed upwards in our solution, the work done by the unknown was positive for an upward Δ.

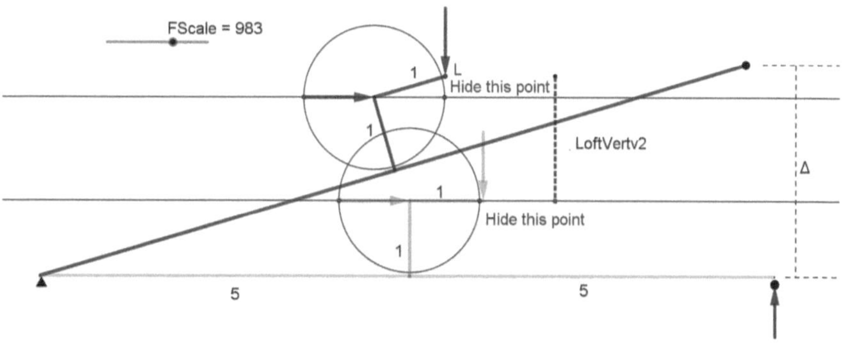

3.11
Hide, rather than delete, extra items

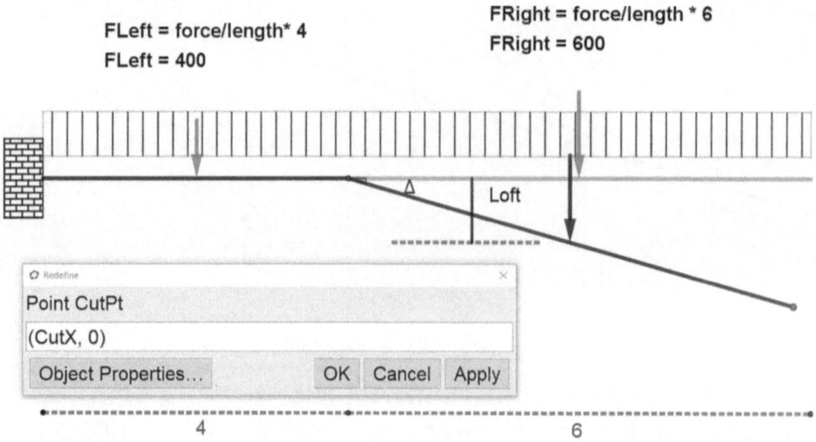

3.12
Insights into internal moment solution

To gain more insights into the geometric solution of internal moment, consider the cantilever beam previously shown in Chapter 2, Figure 2.27. In that problem, the moment at a cut some known distance from the left fixed support was sought. The structure is shown in Figure 3.12, and the parametric location of the cut point, conveniently named "CutPt" (with a Capital C) has a parameter CutX as the X value in Cartesian Coordinates. CutX was previously the input in the Input Bar as CutX = 4, i.e. simply as a number.

There is a length of the beam which swings through a circular arc to the right of the cut. This overhanging length is established and is maintained parametrically, never by means of "hardwiring". For example, for a cut of 4 units from the wall and a 10-unit long total beam length as shown in Figure 3.13, the reader should never create a circle of radius 6 centered on CutPt. Rather, the reader should establish the radius as Radius1 = *x*(Right) – *x*(CutPt) so that if the location of the Right point, or the location of the CutPt changes, things will be automatically updated in the script. Once the circle centered on CutPt is drawn, a Point is snapped to that circle and a segment drawn from CutPt to the Point on the circle. The angle Δ is quickly found from the drop-down menu Angle (select three points). The reader should notice that Δ is immediately updated as the point on the circle is moved. The midpoint between Right and CutPt must be found, the easiest way to do that is to use the MidPoint or Center menu, which is under the Point menu. Notice that we only really need that midpoint, as the load left of the CutPt is not lofted. Having the previously defined

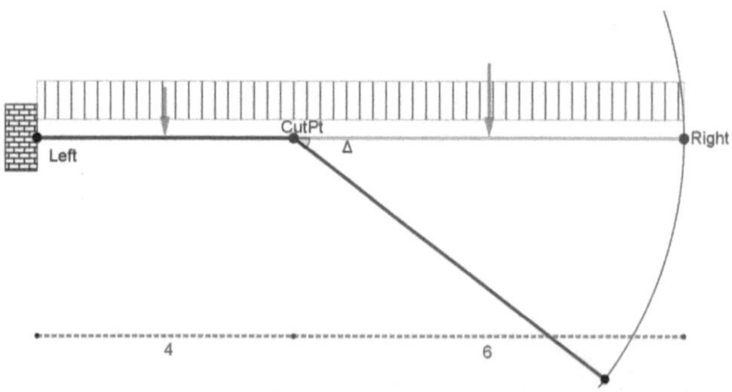

3.13
One cut causes a hinge to form

3.14
Parametrically controlling key terms

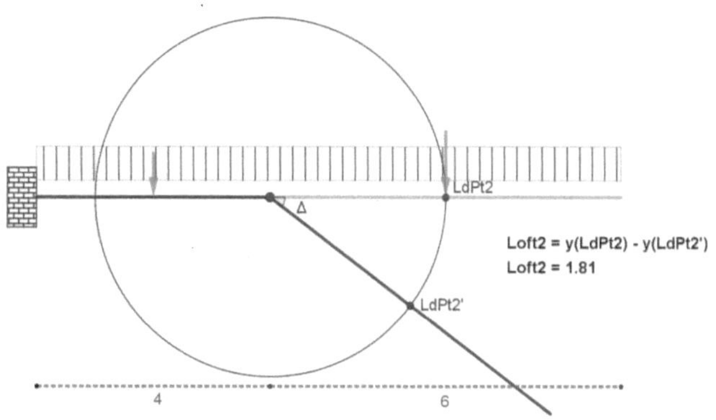

Radius1 is very handy when calculating the magnitude of the concentrated load that is to the right of the CutPt. If the load is input as some variable such as fpl, standing for force per length, then inputting fpl = 100 leads to inputting FRight = fpl Radius1, where a space implies multiplication. Alternatively one could explicitly type a star for multiplication as in FRight = fpl*Radius1 to get 600 units of force.

Establishing the vertical loft could be done as in Figure 3.10, namely either as the difference of the y coordinates of two points, or as the length of a vertical segment. Yet as stated many times previously, the Modern Müller-Breslau requires all elements to follow a circular path in their perturbation. Sines and cosines are not the preferred methods of locating the perturbed load, rather, it is far simpler to draw a circle whose radius is the distance from the cut to the load in the original beam. Then, the load must be on that circle in the perturbed configuration. Once that load's perturbed position is known, it is straightforward to calculate the loft. All these steps are summarized in Figure 3.14.

Note in Figure 3.14, the use of so-called "DynamicText". Text is input anywhere, via the button under the Slider menu icon, second to the right along the top. But Dynamic Text is extremely useful because it can display pre-defined objects from a drop down menu, or an "empty box" can be used to manipulate entries algebraically and in real time.

Any parametric script can be written in many different ways, some are more efficient than others, some simply make more sense to the author. The techniques shown in this textbook do not make any claim to being the best, or the most elegant, they are simply suggestions. Readers may well find better ways of achieving the same results. In fact, one of the joys of being a professional is learning from others and sharing ideas via conferences, publications, or informal discussions. These informal meetings often provide helpful insights to overcome the inevitable stumbling blocks encountered when using any new tool. An example of an alternate way of creating the hinged crack in a beam to find internal moment is shown in Figure 3.15. In Figure 3.15, the height of the crack is controlled by a circle whose radius is placed on a slider and is conveniently named HtCrk. A line perpendicular to the beam is passed through the midpoint of the beam. The circle, whose radius is HtCrk, is centered on that midpoint. The intersection of the line and the circle is found, and the lower intersection is ignored and hidden, not deleted. The angle Δ is then established and that angle is used in Equation 1.1 of Chapter 1.

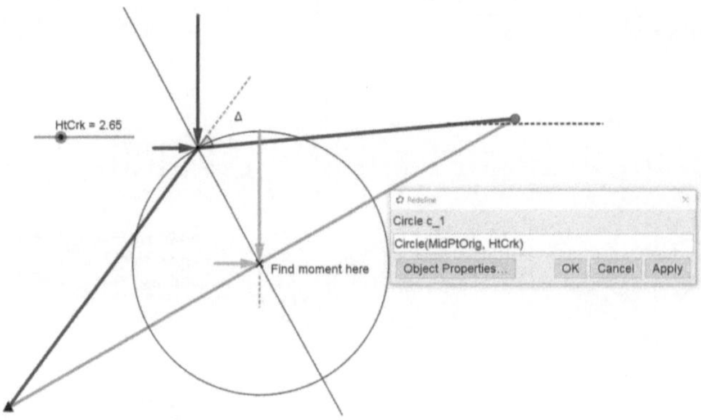

3.15
Alternative way of inducing cracked angle Δ

Recall one of the fundamental rules of the Modern Müller-Breslau, as it applies to investigating moment at a cut of a beam. Namely, that the single cut beam is stretched on either side of the cut. If other members are connected to the ends of the beam being studied, those elements do not change length. Only that single element is elongated. As shown in Figure 3.15, what this requires of the roller on the right is that it remains in its original position, it does not "roll" along the horizontal plane during the perturbation.

CHAPTER 3 ÉTUDES

The following studies or études should be done in GeoGebra.

Étude 3.1) Create a beam that has a roller support at its left end and a pinned support at its right end. But the beam need not be horizontal! Start by defining a total length of a beam and call it LTot, for example, type LTot = 12 in the Input Bar. Create a slider called XPin which spans from a lower limit of 0 to an upper limit of 2*LTot. Create the right node as Right = (XPin, 0), noting that all points must begin with a capital letter. Make the Right node a large triangular symbol using a right click and Object Properties. Next, create a beam that is LTot overall in length by drawing a circle centered on the pin, of radius LTot, a previously defined parametric variable. Draw a point anywhere on the circle and rename it to Roller. Draw a thick segment between the points Pin and Roller. Move the point Roller around to see the circular path taken by the entire beam. A solution is shown in Figure 3.16.

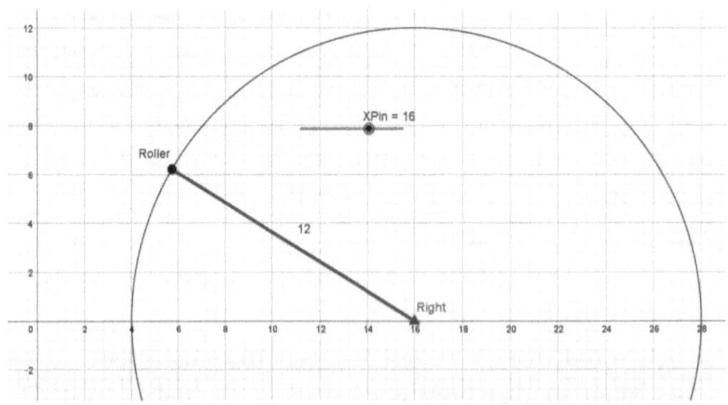

3.16
Étude 3.1 solution

3.17
Étude 3.2 solution

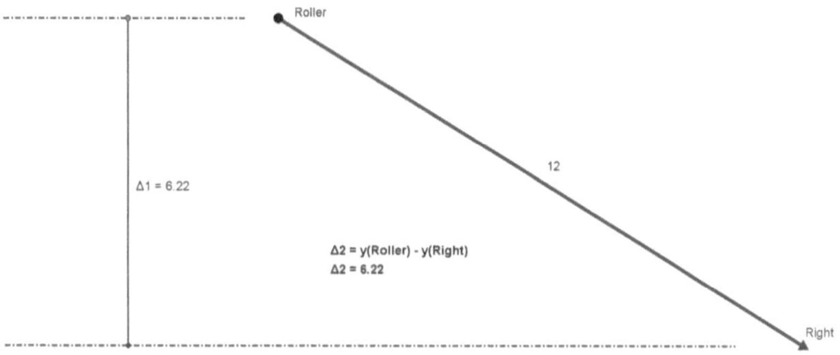

Étude 3.2) Use the script from étude 3.1, but now find the vertical distance of the roller from the horizontal $y = 0$ axis. Do this two ways. Once by creating a vertical segment whose end points align with the Roller and the horizontal $y = 0$ axis. Then a second way, by capturing a variable Loft = y(Roller). Use Dynamic Text to display the variable Loft and change the Object Properties to display the Value (i.e. the Length) of the vertical measuring segment. Hide every extraneous entity. Hide the grid and the Cartesian axes. A solution is shown in Figure 3.17.

Étude 3.3) A simply supported beam 12 units long has a 4 unit long appendage which springs off at 45° clockwise from the beam at the beam's midpoint. This is shown in Figure 3.18. Recreate Figure 3.18, but do so as parametrically as possible. The only variable that is "fixed" or "hardwired" is the Left point at (0,0). Establish a parameter LTot for the 12 unit beam length. Make the Right point a function of the Left point and of LTot. Find the midpoint using the midpoint icon. Establish a 45° clockwise angle using the Angle With Given Size icon. That drops a point in 2D space, through which a line should be drawn. Then, a circle is used, of radius Append, that establishes the length of the appendage. A solution is shown in Figure 3.18.

Étude 3.4) Start with the structure of étude 3.3, but add an entirely new second beam with the same 45° appendage. Make the perturbed beam move in a circular path anchored on the Right pinned support. Start by creating a circle of radius

3.18
Étude 3.3 solution

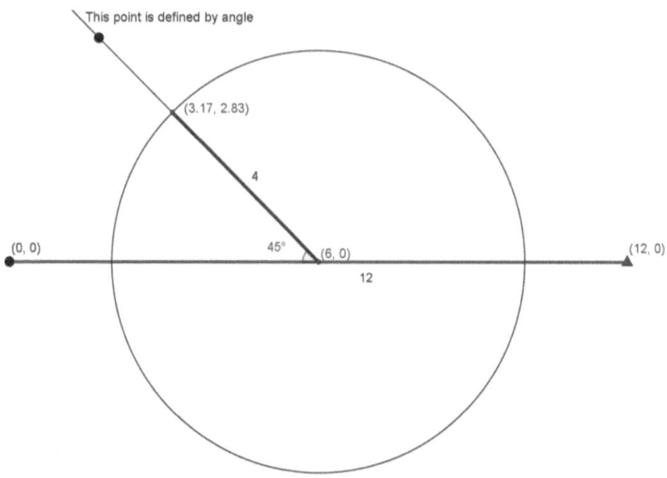

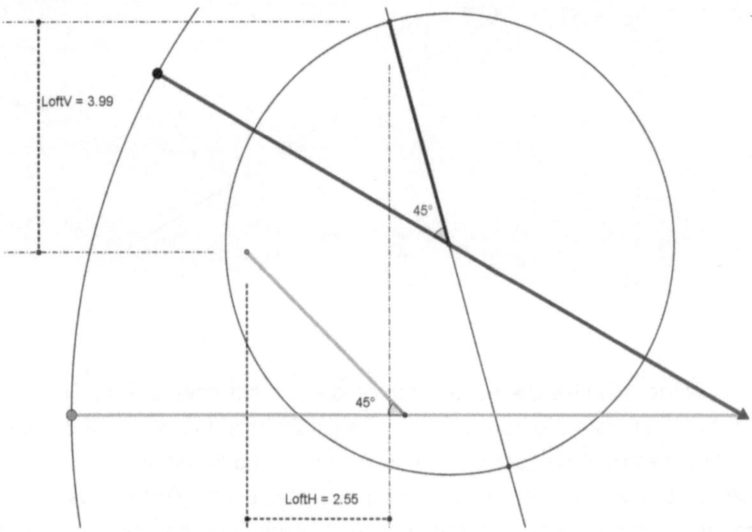

3.19
Étude 3.4 solution

LTot. Drop a new point, renamed as Roller', on the circle. Create a thick segment between Roller' and Right. Change the color of the original structure. Find the midpoint on the perturbed beam. Establish a 45° clockwise angle using the Angle With Given Size icon. That drops a point in 2D space, through which a line should be drawn. Then, a circle is used, of radius Append, that establishes the length of the appendage. Finally, track the vertical and horizontal movement of the tip via a vertical and a horizontal segment, or command such as LoftH = *x*(Roller') – *x*(Roller) and LoftV = *y*(Roller') – *y*(Roller). Display these lofts in your final drawing and remove all extraneous objects. A solution is shown in Figure 3.19 which has many extra objects that would be hidden in a final drawing but are displayed to graphically show the logic of the script.

Étude 3.5) Create a horizontal simply supported beam, pinned at the left end, roller supported at the right end, of total length LTot. Keep everything parametric except the Left pin. Use Fix Object, under Object Properties so as to not accidentally move Left. Create a slider called CutX, with a lower limit of 0 to the upper limit of LTot. Create a force magnitude of some known large number, perhaps Force = 1,000 or Force = 5,000. Recall there are no built-in units in GeoGebra, so keep your units consistent in your own notes. Create a slider called ForceScale, with limits from 1 to Force. Locate a cut point on the beam called CutPt = (CutX, 0). Draw the CutPt using an X symbol. Draw the Force as a vertical vector touching CutPt, the size of the vector is Force/ForceScale, where ForceScale is set to some comfortable number. Add dynamic text stating the magnitude of the downward force. Display the Cartesian Coordinates, but not the names of the end support points. A solution is shown in Figure 3.20.

Étude 3.6) Start with the construction of étude 3.5, but now add a second beam that is lofted due to a variable vertical Δ of the right support. Use a circle, centered on the Left pinned support, of radius CutX, to establish where the load is in the perturbed configuration. Calculate the loft of the load and program the right reaction using Equation 1.1 of Chapter 1. If any number appears as a fraction

3.20
Étude 3.5 solution

3.21
Étude 3.6 solution

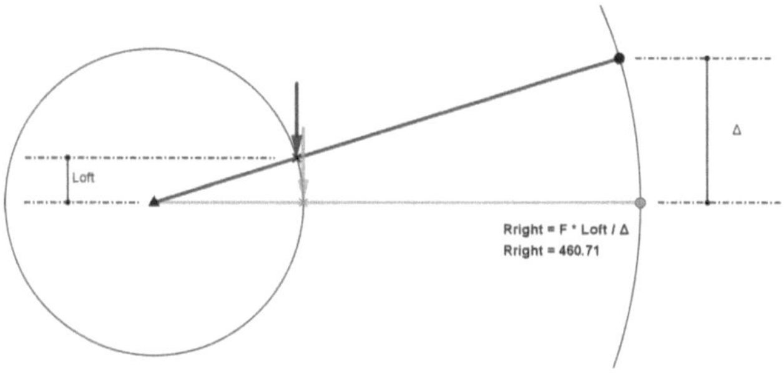

in the Algebra Window, right click, then Object Properties >> Algebra >> Unclick the symbolic box.

Draw at the same FScale as before, the perturbed load as a vertical vector, not as perpendicular to the lofted beam. Use dynamic text to display the answers and make sure the unperturbed configuration is drawn with a different line weight and/or color than the perturbed configuration. A sample solution is shown in Figure 3.21 with two construction circles that eventually would be hidden in a final drawing, but are displayed to show the logic of the script.

Étude 3.7) Start with the construction of étude 3.5, but now add a second beam that is lofted due to a crack in the beam at CutX. Draw a perturbed beam whose relative angle on either side of the cut is called Δ2, as the variable Δ has been used in étude 3.6 already. Find this perturbation by drawing a vertical line through CutX, then place a point on that line, Call that point CutX′. Draw stretched segments from that point to Left and Right supports, respectively. Draw a line from Left through CutX′. Find the angle by either using this new line and a horizontal line at $y = 0$, or by placing a third point on that line through Left and CutX′. Find the loft of the load from étude 3.5, but call it Loft2. Find the moment at the cut using Equation 1.1 of Chapter 1. Display the moment using dynamic text and hide all extraneous objects. A sample solution is shown in Figure 3.22 with several construction lines remaining that would eventually be hidden, but are displayed to show the logic of the script.

3.22
Étude 3.7 solution

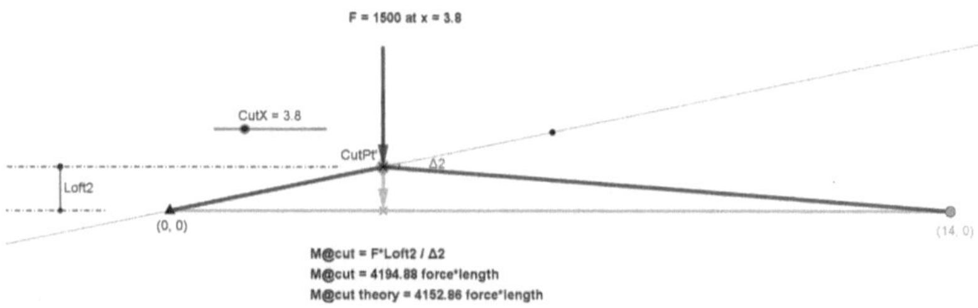

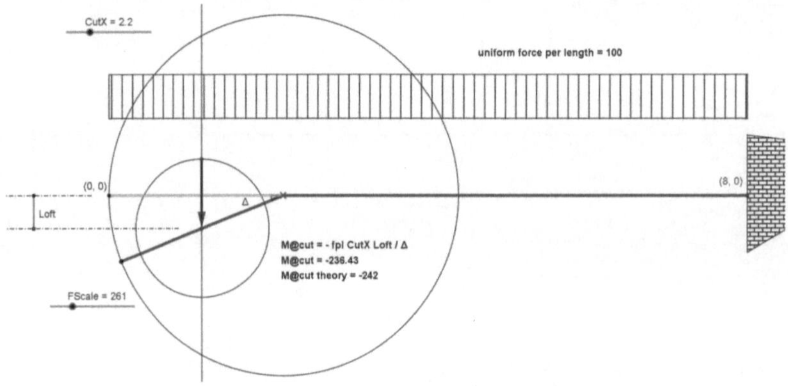

3.23
Étude 3.8 solution

Étude 3.8) Create a cantilever beam, free at the Left end and fixed at the Right end. Assume a uniform force per length is applied to the beam, perhaps fpl = 100. Establish a CutPt somewhere along the length of the beam as was done in étude 3.7. Allow the portion of the beam left of the cut to follow a circular path, and find the loft of the load left of the cut, which will be placed at the midpoint of the perturbed beam segment left of the cut. Then find the moment in the beam at this CutPt using the Δ technique established in étude 3.7. Try a few different Δ values to see if M@cut converges to the theoretical value of $M@cut = -fpl \cdot \frac{CutX^2}{2}$. Be strict with the signs as you apply Equation 1.1 of Chapter 1. Display important information and hide extraneous information. A sample solution is shown in Figure 3.23. A few extra construction lines are left in the drawing to demonstrate the logic of the script. These should be hidden in a final drawing.

Étude 3.9) Create a simply supported, but slanted beam. Establish the two supporting points A1 = (0,0) and B2 = (SpanAB, Span12). Draw a segment to represent the beam between the two points A1 and B2 and call it Beam. Check the length of the segment Beam in the Algebra window, and display it in the drawing, either by using Object Properties of Beam, or by using dynamic text. Establish a CutPt somewhere along the length of the beam. Perhaps the easiest way of doing this is to simply drop a point on the beam and slide it back and forth. Display the Cartesian Coordinates of the CutPt. Drop a second point on the beam and call it LoadPt and place a vertical and a horizontal load, drawn to the same ForceScale, at the LoadPt. Show the magnitudes of the two loads, as well as the Cartesian Coordinates of the LoadPt. A sample solution is shown in Figure 3.24.

Étude 3.10) Start with the construction of étude 3.9, but now find the moment at the cut. Perhaps the easiest way of establishing the lofted cracked beam is to draw a line perpendicular to the slanted beam, drop a point on that line and establish that as the crack as CutPt'. Do not move the roller or the pin and stretch the perturbed beam from CutPt' to A1 and from CutPt' to B2. Remember that Δ is the relative angle of the beam on either side of the crack. A sample solution is shown in Figure 3.25. A few extra construction lines are left in the drawing to demonstrate the logic of the script. These should be hidden in a final drawing.

3.24
Étude 3.9 solution

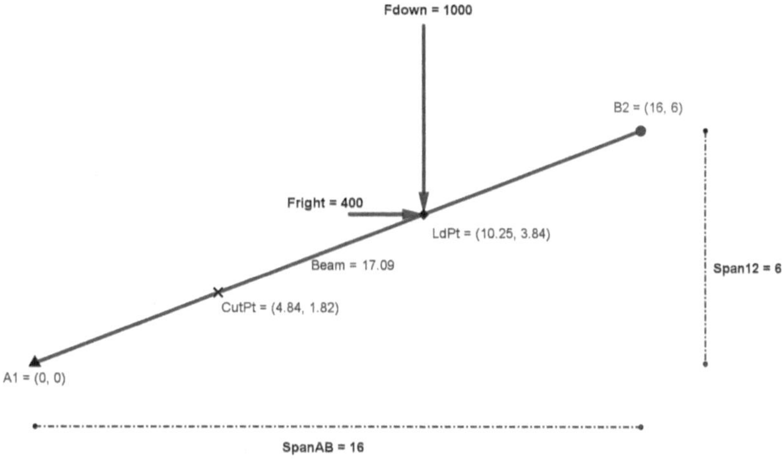

3.25
Étude 3.10 solution

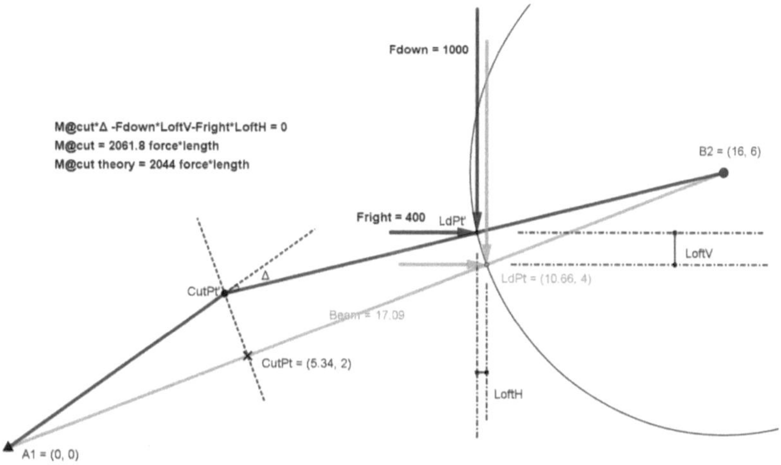

4 Trusses

Two-dimensional trusses are defined as planar structures made up of individual straight elements, known as "bars". Trusses are an unusual subset of structures in that they ultimately act as a beam, spanning long distances, yet the individual elements of the truss are assumed to never bend. It is important to understand that even though the overall structure bends like a beam and becomes concave up typically for gravity loads, the individual bars themselves remain straight and unbent. There are some idealized, mathematical assumptions about classical trusses:

- The individual bar elements are assumed to be "hinged" at their ends, i.e. they cannot transfer bending to the subsequent member. This connection at the ends of truss elements is completely analogous to the "internal hinge" previously shown in Figure 2.23 of Chapter 2.
- The individual bar elements are assumed to be weightless, i.e. they do not bend at all due to their lack of self-weight.
- Any loads applied to the truss may only be applied at the connecting points linking the bars together. These connecting points are called "nodes". Loads may only be applied to the nodes of a truss, never to the bars directly. The reader can imagine that if loads were applied to the bars themselves, those bars would bend, which would violate the fundamental idealization of "no individual bars bend".
- The individual bars experience only pure tension, or pure compression, no twisting, no bending. This is the hallmark of the truss' efficiency; namely carrying load solely through tension and compression, as opposed to bending. This is a very efficient means of load flow, notice how open and airy the trusses of Figure 4.1 appear. Bending of solid beams is an indirect means of load flow, and is far less efficient.

Figure 4.1 shows this curious, somewhat paradoxical idea, of overall beam-like bending, yet all individual bar elements do not bend at all!

An evaluation of any structure requires an informed set of skills. Without such skills, evaluation of structures immediately becomes subjective and the study of trusses draws upon multiple skill sets. Trusses have been used since the Renaissance era, but only in the 1820s and 1830s did they mature and become engineered mathematically. Trusses also allow for a deep-dive study because they were formed by craftspersons, studied by theoreticians, yet their maturation came about

DOI: 10.4324/9781003180913-4

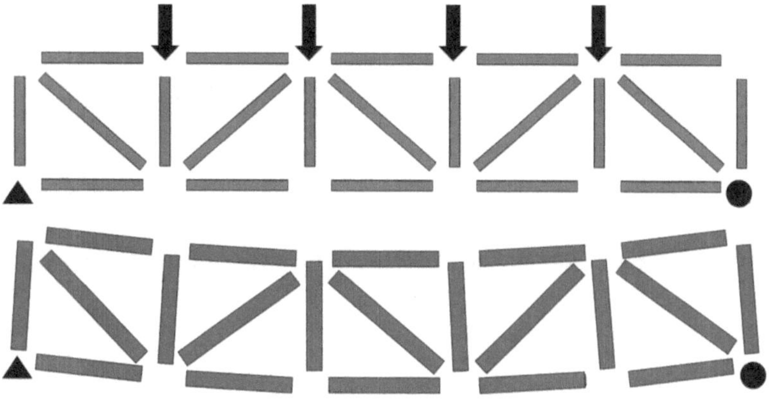

4.1
Truss acts as a beam but all elements remain straight

because of intercontinental travel by inquisitive engineering professors! Sadly, many engineers today are unable to confidently share this story with the general public, because of the nearly complete lack of training in history in traditional structural engineering curricula. Furthermore, engineers and architects both have been trained to look at trusses from a very practical, pragmatic, and prosaic viewpoint. Often, what is lost because of this purely utilitarian viewpoint, is the wonder of how a pin-ended mathematical model found its way into a prolific series of competing patents in the early nineteenth century. The story of trusses is really a microcosm of the wider story of Civil engineering, devoted to "civilian" projects such as bridges. Bridges play a large part in the historical study of trusses because the large spans and heavy loads of trains crossing rivers necessitated an ever more efficient structural system. Ordinary beams could never sustain a locomotive over a 50-meter span. The story of truss developments over time is deeply linked to the economic and social issues surrounding locomotive travel and commerce.

Many people associate trusses with the well-established triangulation of elements, perhaps they recall seeing such triangulated beam-like spans beneath a long bridge. The triangulation of elements took time to develop when looking at the history of trusses. Early truss structures were essentially arches with some truss bars. This textbook will study arches in detail in Chapters 7 and 8. For now, notice how structural designers became more confident with their designs of trusses as time progressed, and as mathematical models of truss behavior became more refined and more widely used. Figure 4.2 gives a quick overview of the obvious reliance on the arch in 1765 with the European Grubermann Bridge. Then, in the United States, rapid improvements in truss modeling led to a patent on the Burr Truss in 1832 shown in Figure 4.3, and a patent on the Long Truss of 1830 shown in Figure 4.4, where the arch practically disappears.

As structural designers continued to gain confidence, more and more innovative truss shapes developed. These new shapes, developed in the nineteenth century, eventually eliminated arches and solely relied on tension/compression of the individual truss bar elements. Notice in Figures 4.2–4.4 that real structures were shown, which are clearly three-dimensional, yet a two-dimensional "elevation" drawing or a frontal view, is sufficient to understand and to model the load flow. This is an important point for the reader to grasp, namely, that two-dimensional

4.2
European-style eighteenth-century truss, the Grubermann Bridge 1765

models can indeed accurately capture the behavior of real structures. Yet sometimes, the structures have elements that simply extend into the third dimension. We end this chapter with a study of just such trusses, called "space trusses" or "spatial structures".

In his 1982 book *Horizontal Span Building Structures*, Professor Wolfgang Schueller elegantly describes the classification of two-dimensional trusses into distinct typologies. Schueller presented trusses as falling into several categories, or typologies.

- Simple Trusses, which strictly adhere to the familiar triangulation pattern.
- Irregular Trusses, which have long members who pass over other members without a joint or "node" at the crossing point. Note that the trusses in Figures 4.3 and 4.4 all have members which pass in front of each other, or behind each other, without creating a node at the intersection.
- Trusses as Beams, where one bar at the top or bottom of the truss is perfectly horizontal. This extreme top or bottom bar of a truss is referred to as a "chord" member, whereas the interior bars enveloped by the chords are known as "web" members.
- Trusses as Frames or Arches, where enormous spans are crossed. In the arch configuration, the truss rises to some peak, this peak is known as the "crown". In the frame configuration, the truss has a beam-like portion in the central span and column-like supports which elevate the beam.

Figures 4.5–4.8 shows examples of each of these four truss typologies. In Figures 4.5 and 4.6, a light wire shadow displays the original, unloaded configuration of the truss. Then the darker lines show how the truss deforms under the applied loads. All trusses in Figures 4.5 and 4.6 are subjected to identical gravity loads, one load on each of the interior nodes on the bottom chord.

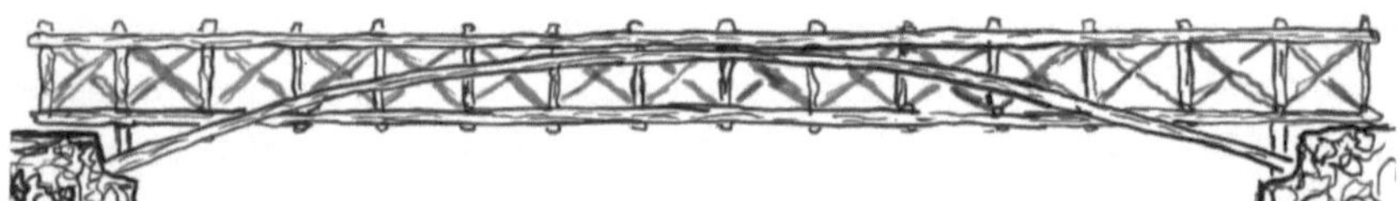

4.3
Burr Truss Bridge, patented 1832

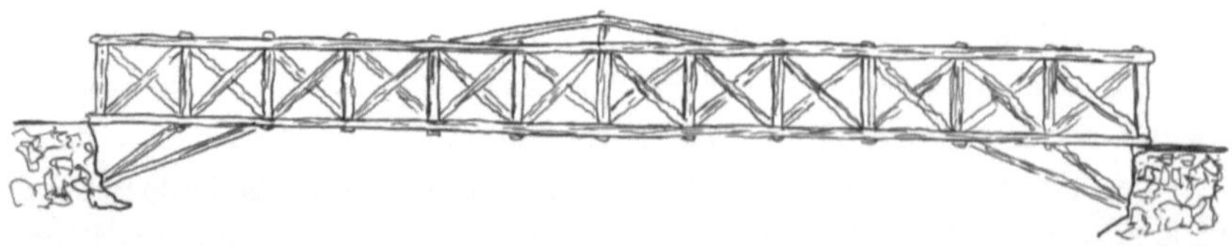

4.4
Long Truss Bridge, patented 1830

When studying Figures 4.5–4.8, the reader should check that each of these structures are classified as "determinate". Recall the definition of "determinate from Chapter 1:

> For structural elements, classified as "determinate", the external equilibrating reactions are unambiguous and do not depend on the material properties of the element, nor do they depend on the cross sectional properties (round, square, rectangular) of the elements. This is what is meant by "determinate", i.e. the reactions are unambiguously determined. For structural elements classified as "indeterminate", the external equilibrating reactions are not immediately found because the material properties of the element, as well as the cross sectional shape do indeed come into play.

In the study of beams, it was quick to establish whether or not a beam was determinate. In two dimensions, typical determinate beams have three unknown external reactions, typically two vertical reactions and one horizontal reaction. For gravity loads only, the horizontal reaction is zero for horizontal beams. Furthermore, in some statically determinate beams, the introduction of an internal hinge necessitated the addition of one external reaction, and this was necessary to prevent a collapse mechanism from forming. These collapse mechanisms are very important to structural engineers. They will be studied in Chapter 10 via so-called "Plastic Analysis". Returning to determinacy of trusses, there is a simple equation that checks if a truss is determinate. The reader should verify that the following Equation 4.1 is equal to zero as a check of determinacy in the examples that follow.

$$b + R - 2 \cdot j = 0 \qquad \text{(Equation 4.1)}$$

Where "b" is the number of bars, "R" is the number of external reactions, and "j" is the number of joints, or "nodes" which include the support joints on the ground.

All of the trusses shown in Figures 4.5–4.7 satisfy this equation.

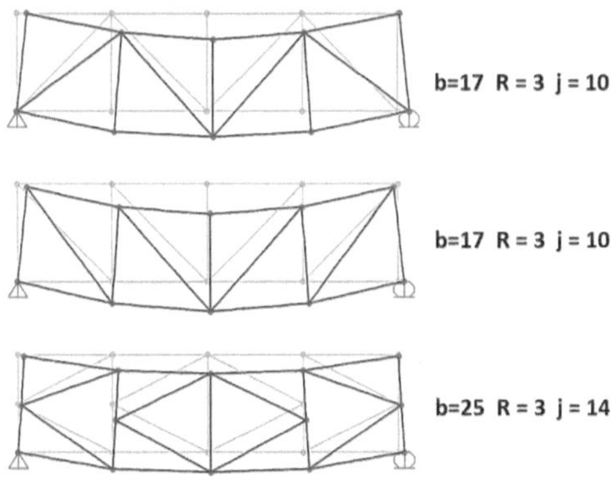

4.5
Three "simple" trusses subjected to identical downward loads

As before in Figure 4.5, in Figure 4.6 all trusses are subjected to identical gravity loads, one load on each of the interior nodes on the bottom. Note there are no bottom chord members in the third truss of Figure 4.6. The reader should check the determinacy equation to ensure that all these are indeed determinate.

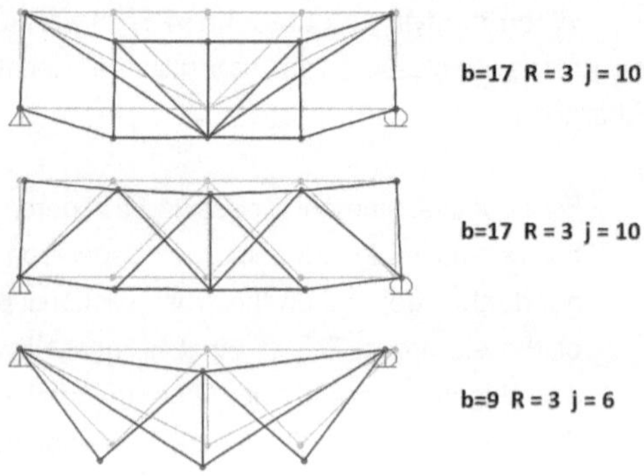

4.6
Three "irregular" trusses subjected to identical downward loads

Figure 4.7 shows "beam-like" configurations, where the designers made the truss deeper in the center of the span and less tall near the supports. Again, the reader should check the determinacy equation.

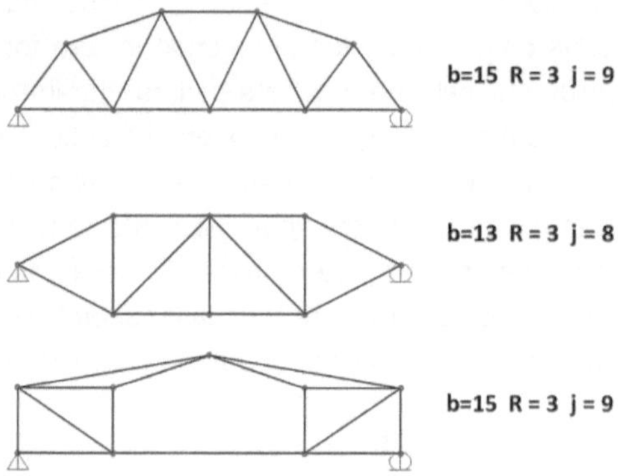

4.7
Three "beam-like" trusses in their undeformed configurations

Figure 4.8 shows three "arch-like or frame-like" trusses, which are all loaded the same amount with downward loads. The diagram of Figure 4.8 is a qualitative depiction of the magnitude of axial internal force in each element. The width of the lines in the map of internal forces reflects the severity of axial load in any particular bar element. Of course, this depends on the applied loads, so for now the reader should simply recognize that for any given externally applied load on a truss, the individual bar members of the truss carry loads axially, some have large internal axial forces, others have small internal axial forces. Some bars will feel tensile forces, other bars will feel compressive forces. It is clear in the middle truss of Figure 4.8, that the axial forces are significantly larger than in the other two trusses. This demonstrates the power of the arch and the weakness of the beam as a means of spanning horizontal distances. It is these internal axial forces that we can establish with the Modern Müller-Breslau Method, along with the external equilibrating reactions at the ends of any truss. Finally, the reader should check the determinacy equation for these three trusses.

It is important to reinforce to the reader that external reactions of any truss can be found using the Modern Müller-Breslau Method in the exact same way as the external reactions of a beam were found. In many determinate trusses, there will

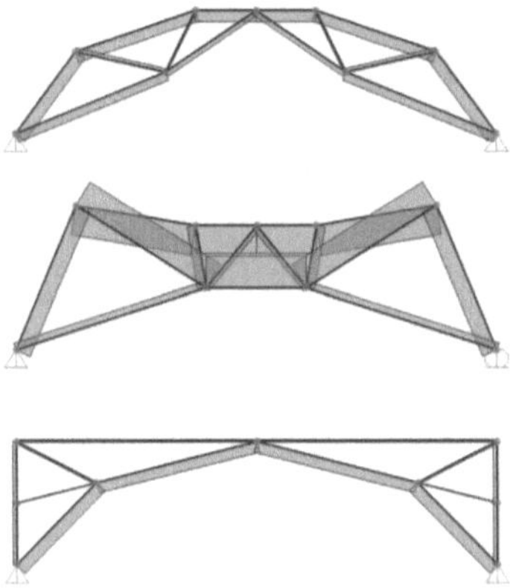

4.8
Three "arch-like" or "frame-like" trusses with a map of qualitative axial force magnitudes

be three unknown external equilibrating reactions, exactly like many determinate beams. In some trusses, for example as those at the top and at the bottom of Figure 4.8, there are four unknown reactions. This is analogous to the beam with an internal hinge that was studied previously. Note that at the crown of the trusses at the top and at the bottom of Figure 4.8, there is indeed an internal hinge, necessitating a fourth external equilibrating reaction.

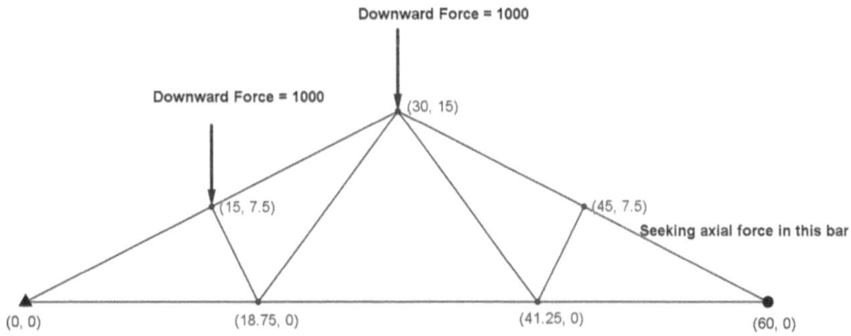

4.9
Example truss, seeking a particular bar axial force

To begin unlocking the internal forces of any bar in a truss, consider the structure shown in Figure 4.9. The Cartesian Coordinates of each node are shown, with consistent units of length. Two downward loads are applied to the top nodes of the truss, and we seek the axial force in the top chord on the far right side.

The reader may be surprised that the technique is exactly the same as it was for any other problem in this textbook, and we still only use Equation 1.1 of Chapter 1. That Equation 1.1 is restated here as Equation 4.2.

$$Unknown \cdot \Delta + \sum Force_i \cdot loft_i = 0 \qquad \text{(Equation 4.2)}$$

This Equation 1.1 is restated now for the specific purpose of establishing an internal bar force. The Unknown is the bar force and the Δ is a stretching or a shortening along the length or "longitudinal axis" of the bar. The work done by the Uknown Bar Force is always negative for any perturbation Δ. This will be described at the end of this chapter in more detail in Figure 4.18, but the reason is that for any Δ, the resulting internal axial force must always oppose Δ. Thus, Equation 1.1 of Chapter 1 can be restated here as Equation 4.3, which applies to any particular axial bar force.

$$-\textit{Unknown Bar Force} \cdot \Delta + \sum \textit{Force}_i \cdot \textit{loft}_i = 0 \qquad \text{(Equation 4.3)}$$

The internal work of a perturbed bar element will always be negative as shown in Equation 4.3.

In the truss of Figure 4.9, the Unknown is the axial bar force in the top chord member on the far right side of the truss. The Δ then must be a perturbation in the direction of the Unknown, which we perturb after "removing" the ability of the structure to carry the Unknown. Here in the truss, that means that the bar of interest is cut, and an axial displacement Δ is applied along the axis of the Unknown, i.e. in the direction of the Unknown as was done in all the previous examples.

Figure 4.10 shows the quantitative solution for this bar force, in a grossly perturbed truss. The reader should take careful notice of the fact that the perturbed bar has dramatically changed orientation, i.e. that the final perturbation Δ is not along the original top right chord axis! For some trusses such as the one shown in Figure 4.10, the entire truss distorts because of a single bar perturbation. Yet, later we will also investigate a different truss, where the perturbation distorts only one portion of the truss and leaves another portion unperturbed.

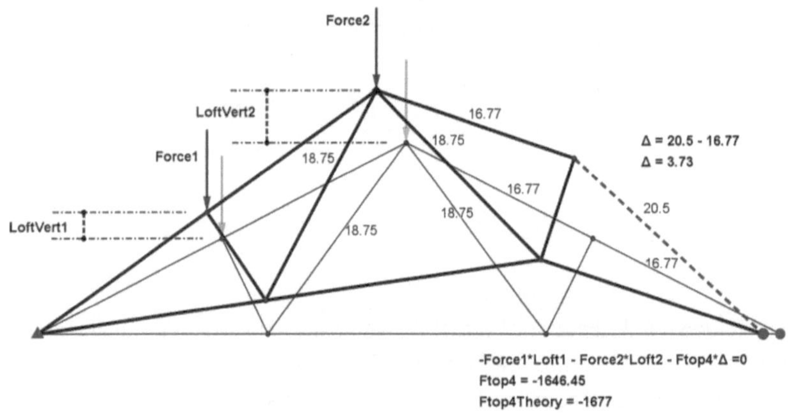

4.10
Example truss, solving for a top right chord bar axial force

The steps of how to draw Figure 4.10 in GeoGebra will be described in detail in Chapter 5. For now, the reader should explore deeply the fundamental idea of a globally lofted truss induced by a single bar's axial perturbation. Figure 4.11 shows an elementary truss, with three bars. The truss is resting on a known surface, shown as a pier on either side of a bridge span. The truss is subjected to two loads and the bottom chord bar force is sought. The original lengths of each bar are also shown.

Since the lower bar force is the Unknown of Equation 1.1 of Chapter 1, it is the one that is cut and perturbed, i.e. elongated an amount Δ. In Figure 4.12, Δ = 1

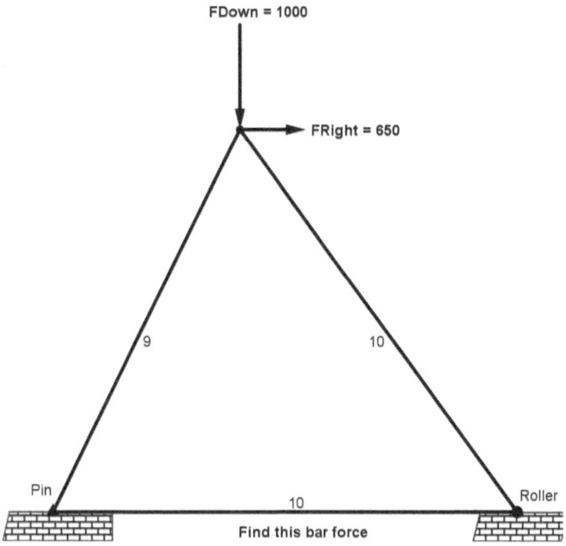

4.11
Elementary truss, solving for a particular bar axial force

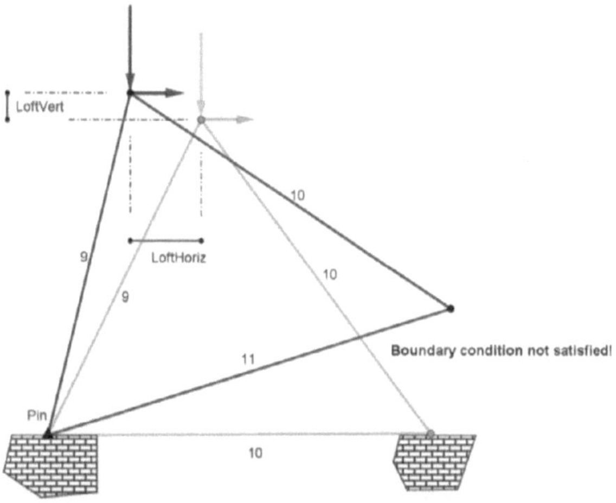

4.12
Reasonable Δ = 1 but enormous boundary condition violation

which is fairly large, but note that the boundary condition violation is enormous. What is meant by "boundary condition violation"? It is the fact that the lower right roller is nowhere near the horizontal support, where it must be. That roller must not move vertically, but since it is a roller, it can be allowed to move horizontally. Notice that two of the bars did not change length, but the bar whose force we are seeking is elongated by $\Delta = 1$. This is a critical concept that will be repeated several times.

Figure 4.13 shows how the boundary condition is manually adjusted to be nearly satisfied, yet the Δ remains very large. The reader should understand this step by experimentation with sketches or with elementary strips of paper or pieces of wood.

Even for this large Δ of Figure 4.13, the final answer is quite close to the theoretical value. Figure 4.14 shows the answer for a small Δ and for a small boundary condition violation. This final answer approaches the theoretical value.

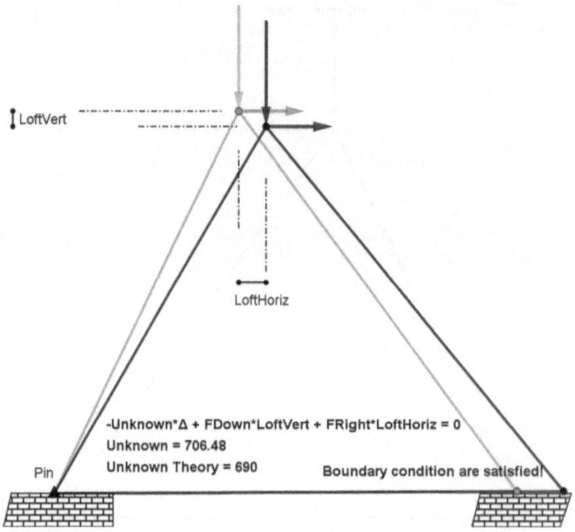

4.13
Reasonable Δ but very tiny boundary condition violation

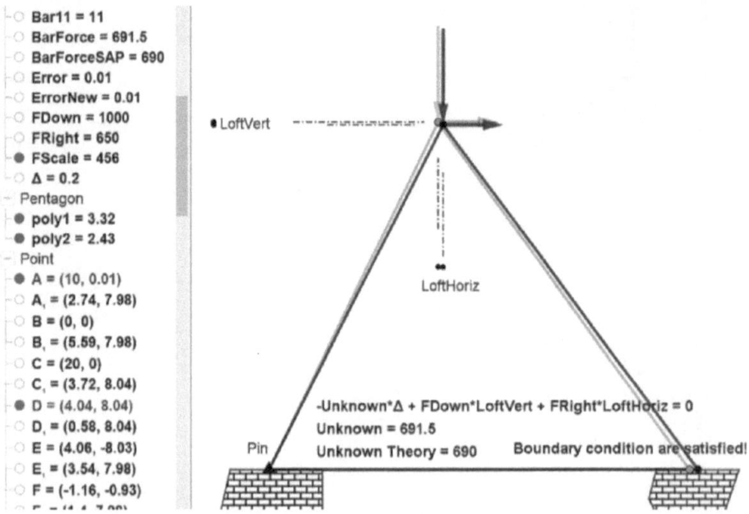

4.14
Small Δ and small boundary condition violation

Consider now the third truss of Figure 4.5. Suppose it has three downward loads and one lateral load to the right as shown in Figure 4.15. Note that there are three unknown reactions, although it is clear that the horizontal equilibrating reaction at the left pin support must be equal to the lateral load but in the opposite direction. Suppose we sought the lower right web bar axial force, and keep in mind that the original, unperturbed length of this element is 13.4 units of length long, i.e. $\sqrt{12^2+6^2}$.

The solution for an enormous perturbation is shown in Figure 4.16. The reader should study Figure 4.16 and notice that the only member that changed length is the unknown member whose force we are seeking, the one highlighted by "Force?" in Figure 4.15. All the other members maintained their original length. But also notice that the roller boundary condition was grossly violated in Figure 4.16!

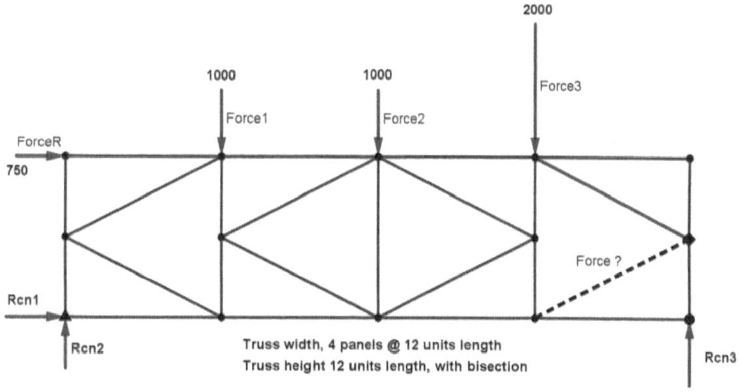

4.15
Setup for complicated regular truss

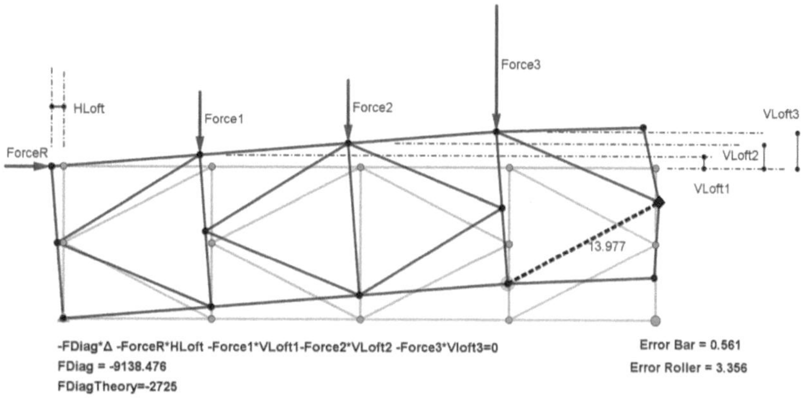

4.16
Solution with grossly distorted truss

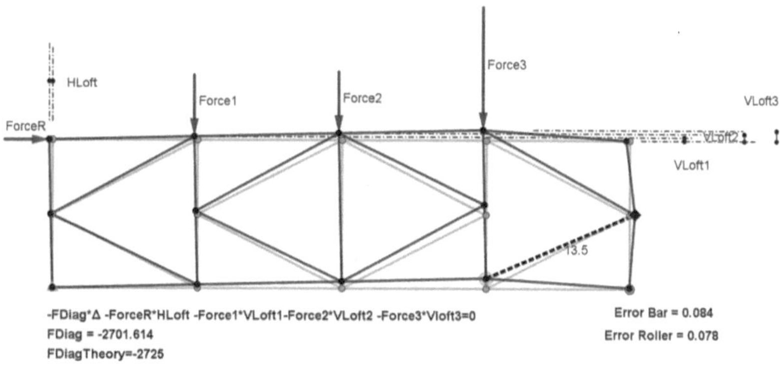

4.17
Solution with small perturbation along unknown and small boundary condition violation

The GeoGebra constructions of these trusses will be described in detail in Chapter 6. For now, the reader should look at Figure 4.17 and notice that the perturbation along the stretched diagonal is very small, what started at 13.4 units of length is now 13.5. The boundary condition violation of a zero vertical movement on the lower right support of the truss is also very small. This leads to an answer for the bar force that approaches the theoretical answer.

The Modern Müller-Breslau Method can indeed be used to solve for any external equilibrating reaction or any internal bar force of a determinate truss. But admittedly, the process is a bit involved because of the linkages which must be created, ensuring that only the bar with the Unknown internal force has a change in length. Yet, the power of the method will truly shine when certain three-dimensional trusses are studied later. A powerful feature of the method is that it becomes immediately apparent which bars are engaged and participate in the load flow of a structure, and which do nothing at all to carry a given load. This is clear in two-dimensional as well as in three-dimensional models.

The reader may have questioned why the first term in Equation 1.1 of Chapter 1 must always be negative when applied to a truss bar force. The answer is that if Δ is an elongation, the internal bar force must resist the pulling away of one end of the bar, thus Δ opposes the Unknown force. Conversely, if the bar is contracted or shortened by some Δ, the internal bar force must resist the shortening by pushing on the moving end of the bar, again Δ opposes the Unknown force. This is demonstrated in Figure 4.18 which shows a cantilever beam element that has two loads applied to its free tip, and the cantilever is tilted at a known 55° to the horizontal. While the cantilever is not a truss bar element as it has bending, an internal axial force is still present in such elements and it will serve to demonstrate the negative work of Equation 1.1 of Chapter 1.

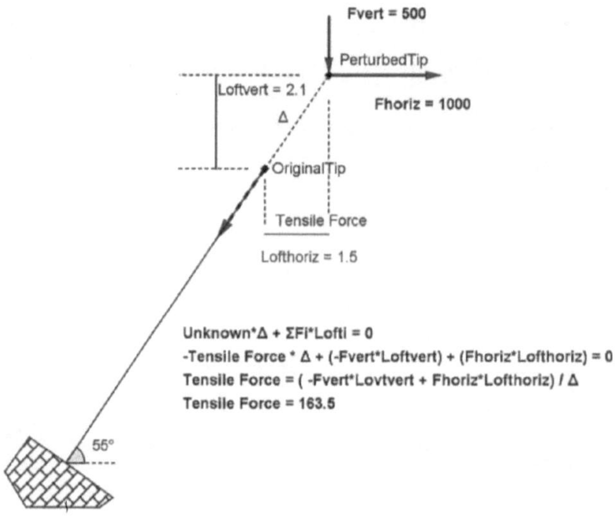

4.18
Internal axial work is always negative at setup

All of the exact same principles described for two-dimensional, planar trusses apply to three-dimensional, so-called "space trusses", or spatial structures. The prestigious international professional society IASS, the International Association of Shell and Spatial Structures still uses the term "spatial structures" as a recognition of the heroic forms created by master designers. Those luminaries include such people as the IASS founder, the Spanish professor and engineer Eduardo Torroja, the brilliant Spanish/Mexican architect Felix Candela, and the legendary Irish engineer Peter Rice and many, many others. We will only briefly look at the simplest three-dimensional spatial structures, such as a tripod and a space truss made up of only a few members. The use of a parametric drawing program makes the study of such structures much more approachable and quantitative than even the finest analog sketches in a notebook could achieve.

Consider the simplest of all spatial structures, a pin-ended tripod truss. Suppose a single downward force was applied to the space truss shown in Figure 4.19 and that we sought the force in one bar element, here labeled Bar1. The method is exactly the same as in two-dimensional trusses, namely, that the single bar being investigated must experience a perturbation Δ along its axis. In three dimensions, it is important to draw a line along the longitudinal axis of the member being investigated, and then to stretch it along that axis. The difference between the original length and the final length is Δ, and the work done is always negative, because as before, for a stretch, the bar pulls back. In other words, the work in the bar is always negative, regardless of Δ. Then, since the other bars are not elongated at all, they must be used to find the new Crown, or meeting point. How they do this is an interesting story because that story tells us of the stability of sub-structures, i.e. how to identify unstable mechanisms in members. That will be shown soon in this chapter, and it is different than the Chapter 10 buckling problems, which are also referred to as "stability studies". In Figure 4.19, since we seek the Unknown force in Bar1, it is Bar1 alone that is perturbed. Here, a Point labeled "MoveMe" is forced to slide along the original Bar1 axis. The reader should be careful to note that MoveMe is not the new Crown! MoveMe is a point that establishes Δ. Now the entire three-dimensional structure must move to accommodate the new length of Bar1. This leads to Crown', the final meeting point of the three bars.

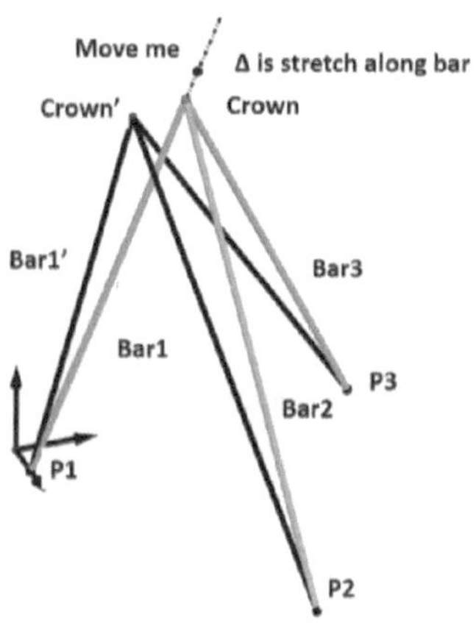

4.19
Qualitative example of bar force in a tripod

The reader should now ask "why is MoveMe not Crown'?" The answer to that question leads to insights into the overall stability of any truss. Consider the same tripod with only Bar2 and Bar3 meeting at the original Crown. Is that structure stable in three dimensions? Is that structure stable in two dimensions? Figure 4.20 answers the first question.

From Figure 4.20 it is immediately apparent that two pin-ended bars joined at a Crown are unstable in three dimensions because a circular mechanism exists, i.e. the two bars would simply flop over if there was any force perpendicular to the plane formed by the two bars. However, in two dimensions, where only gravity

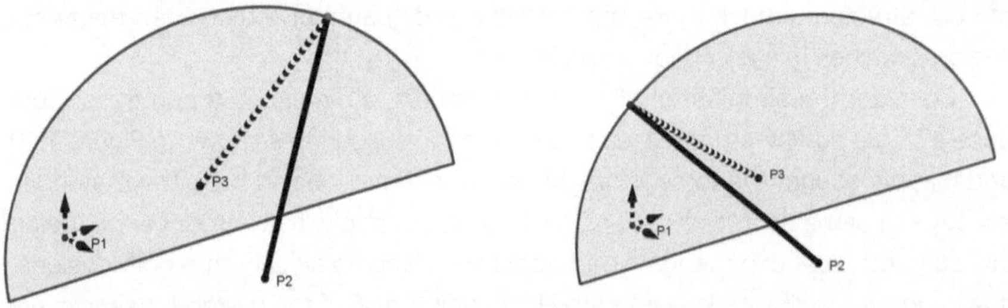

Instability of two truss bars is a circular path

4.20
Two truss bars are unstable in 3D as a circular mechanism exists

loads or horizontal loads exist, as was described in the discussions of elevation views in Figures 4.2–4.4, the two pin-ended bars are stable! In 2D, no forces can be applied perpendicular to the plane of the truss. In 2D, the two bars of Figure 4.20 would form what is known as a "three hinged arch" which will be studied in detail in Chapters 7 and 8.

We can drill down even deeper than Figure 4.20, and question what instability mechanism would exist for a single pin-ended bar in a three-dimensional space? Then, what mechanism would exist for a single pin-ended bar in a two-dimensional space? Figure 4.21 shows the answers to both of those questions.

Single truss bar pinned at one end

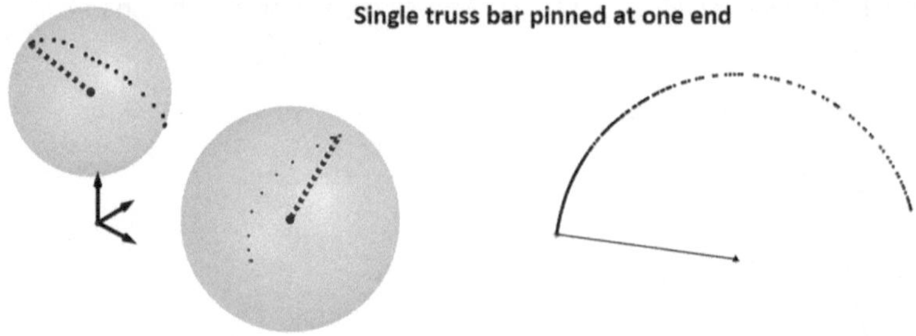

In 3D a spherical mechanism exists　　In 2D a circular mechanism exists

4.21
Single bar pinned at its base is unstable, mechanism is spherical or circular

In 3D, a single pin-ended truss bar would have a spherical mechanism, i.e. if one end is pinned to the ground, the free end travels in a spherical path. In 2D, that same bar's free end travels in a circular path. In both cases, the radius of the path is the length of the truss member itself. Notice another insight, on the left of Figure 4.21. That is, if the base points of the two bars of known length are pinned too far apart from each other, there can be no intersection of the two paths! And if the two spheres do in fact intersect, that intersection is the circular path shown in Figure 4.20. The intersection of two spheres is a circle.

How does this all relate to Figure 4.19 and the difference between Crown' and MoveMe? Consider the problem as having two mechanisms. One mechanism is that of Bar1, namely a spherical path. The second mechanism is that of a triangle formed by Bars2 and 3, which flops over in a circular path. If the bases are close

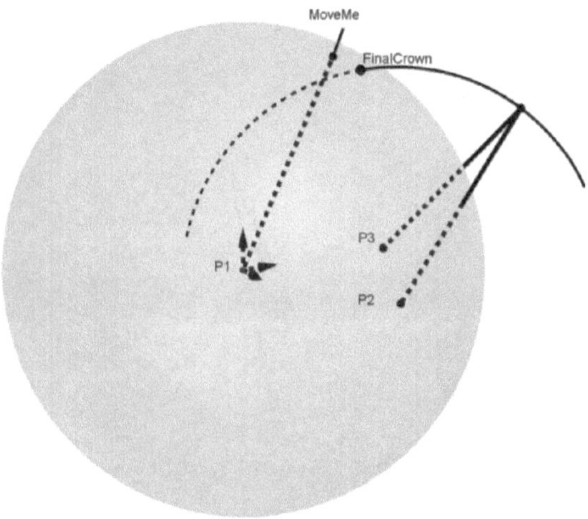

4.22
A sphere and a circle intersect at a single point

enough such that the mechanisms intersect, then the intersection of a spherical path and a circular path is a single point! That point is Crown′. This truth is demonstrated in Figure 4.22.

Compare Figures 4.19–4.22. In Figure 4.22, Bar1 got stretched via the MoveMe point and has a spherical mechanism if we consider it alone. Bar2 and Bar3 remain unchanged, thus they form a triangle that flops over in a circular path. The intersection of that spherical path and the circular path is the final meeting point for all three truss members.

Chapter 5 will demonstrate some techniques for capturing this final meeting point in a parametric drawing environment, as this will be very difficult to do with a hand drawing. However, we qualitatively solve for the Bar1 axial force of Figure 4.19 by once again applying Equation 1.1 of Chapter 1. The loft of the load is the movement it takes, from Crown to Crown′. If there is only a vertical load, then only a vertical loft is measured.

Consider now, extending a stable tripod by adding one more node above the ground level, in 3D space. Figure 4.23 demonstrates the process needed to stabilize the new structure. Points P5 and P6 are on the ground plane. Bar5 and Bar6 form a triangle with the ground, and as such, there exists a circular mechanism, stylized by the path taken by Point C in Figure 4.23. If we stabilize the structure by adding a Bar4 from the Crown1 point, that Bar4 has a spherical mechanism, whose radius is the length of Bar4. One point exists above ground where the circular mechanism and the spherical mechanism intersect, that intersection point is the FinalCrown.

The unperturbed six member space truss is shown in Figure 4.24.

The same principle that was demonstrated in Figures 4.23 and 4.24 is used if we want to establish the force in Bar4 due to some loads. Yet, that same principle also gives a tremendous insight into load flow, in a way that other methods cannot. For example, suppose both Crown1 and FinalCrown are subjected to some known set of loads, horizontal and vertical, and that we sought the axial force in Bar4. We elongate Bar4 by some Δ along its original axis, and we establish a new FinalCrown

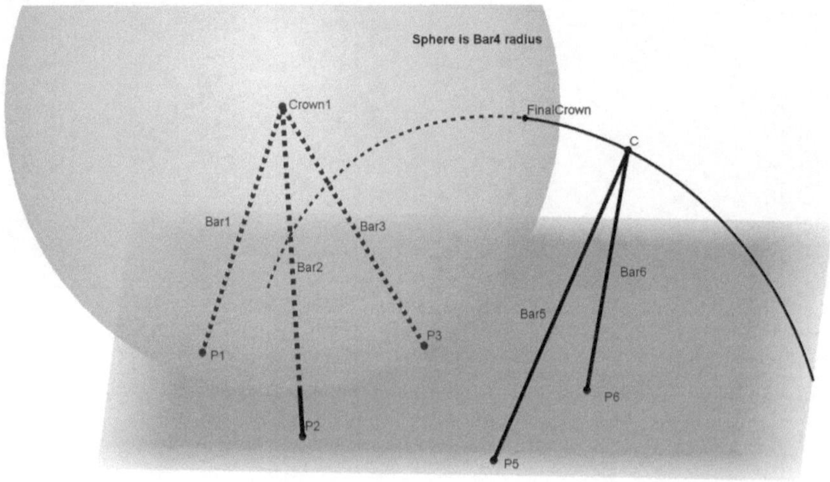

4.23
A sphere and a circle intersect at a single point in the original structure

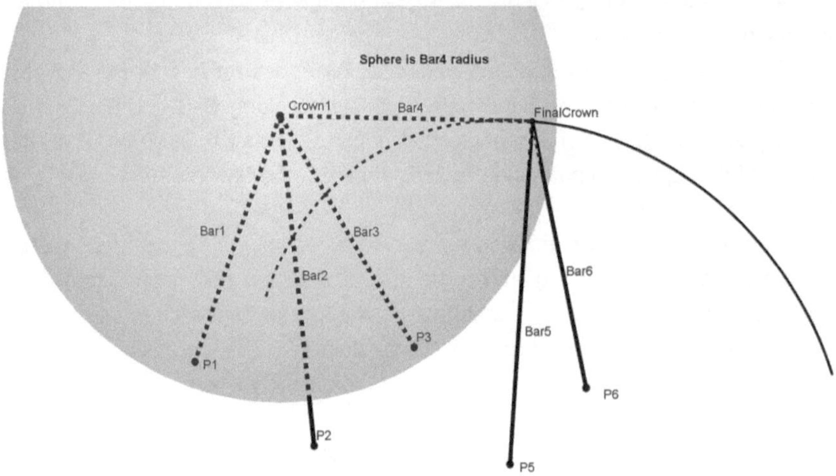

4.24
A sphere and a circle intersect at a single point in the original structure

Point. First, notice that the triangle formed by Bar5 and Bar6 in Figure 4.24 is leaning in closer to the tripod than it does in Figure 4.25, where it leans a bit away from the tripod. This is caused by the perturbation imposed upon Bar4. But now notice the insight! Any loads applied to Crown1 of the tripod cause no axial force in Bar4, only the loads on the FinalCrown Point cause force in Bar4! This is because Crown1 does not move at all due to the perturbation of Bar4. Only the FinalCrown (or Point C of Figure 4.23) moves due to the perturbation of Bar4.

To establish the axial force in Bar4, we already have seen that only FinalCrown moves, thus only loads on FinalCrown cause work to be done and thus contribute to the axial force in Bar4. If there were loads in the X, Y, and Z directions applied to the FinalCrown, then the lofts for Equation 1.1 of Chapter 1 would be the "gap" in the X, Y, and Z directions, where the "gap" is the distance between the FinalCrown and the FinalCrownBefore, i.e. between the shape where Δ is non-zero, and where Δ is zero. This is shown in Figure 4.26.

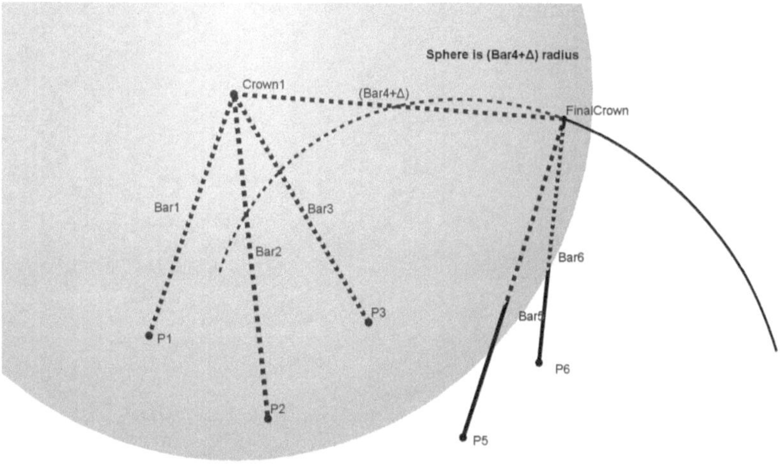

4.25
Crown1 does not move, only FinalCrown moves

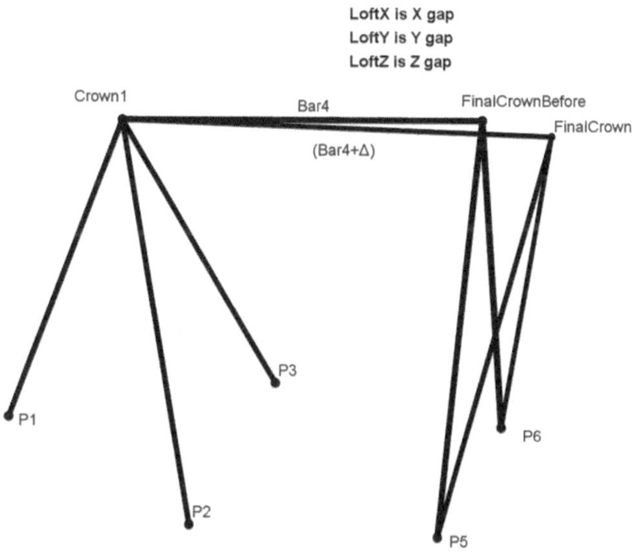

4.26
Break up the gap into *X*, *Y*, and *Z* components

TRUSS ÉTUDES

The following studies or études should be sketched by hand.

Étude 4.1) A truss has three members, the top chords are each 10 units of length and the bottom chord is 12 units of length. The truss is pinned at the left end and is roller supported vertically at the right end. There is a downward load of 500 and a leftward load of 250 units of force applied to the crown. The crown has Cartesian coordinates of (6,8). Sketch by hand the setup shown in Figure 4.27.

Étude 4.2) For the truss of Figure 4.27, find the right vertical reaction. A hand sketched solution is shown in Figure 4.28.

Étude 4.3) For the truss of Figure 4.27, find the bottom member's internal force. A hand sketched solution is shown in Figure 4.29.

Étude 4.4) For the truss of Figure 4.27, find the right member's internal force. A hand sketched solution is shown in Figure 4.30.

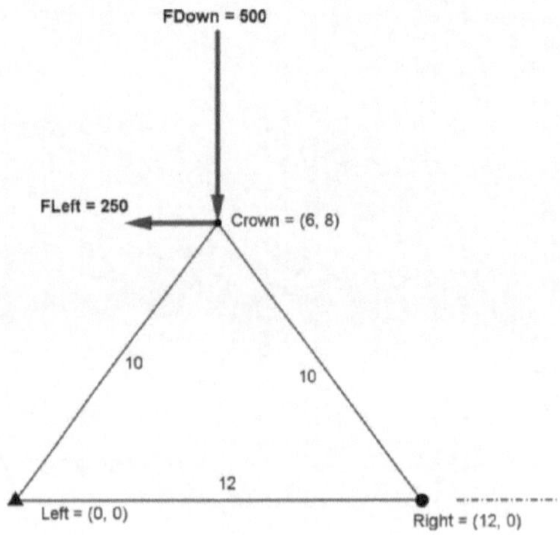

4.27
Étude 4.1 solution

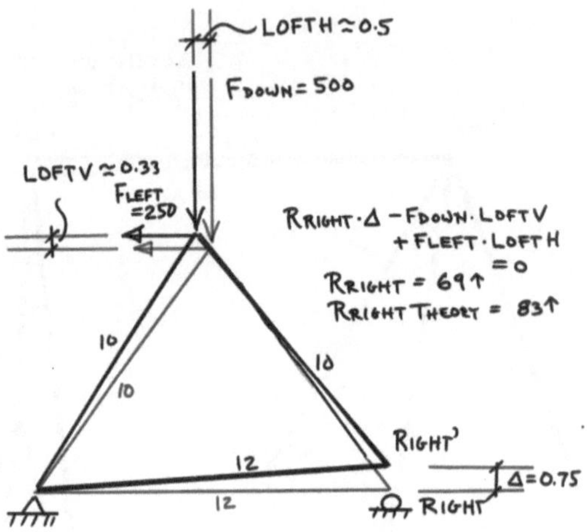

4.28
Étude 4.2 solution

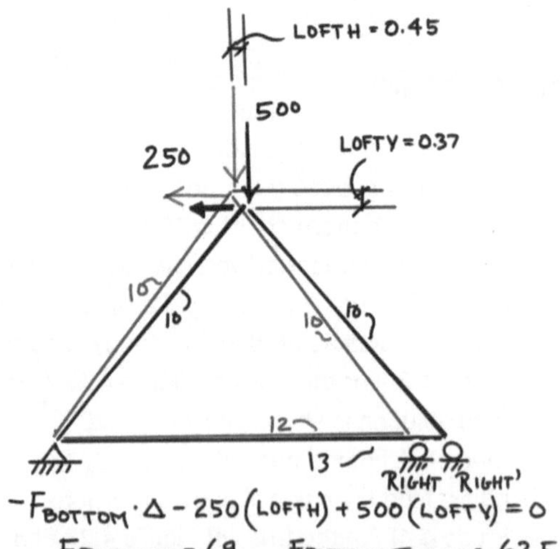

4.29
Étude 4.3 solution

Étude 4.5) The following truss shown in Figure 4.31, has five members and is 15 units wide and 12 units tall, pin supported at lower left and vertically roller supported at lower right. First, check to see if the truss is determinate or not. Then sketch the setup by hand.

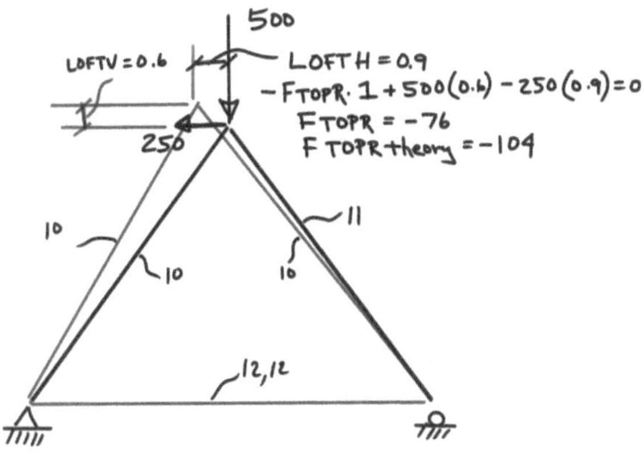

4.30
Étude 4.4 solution

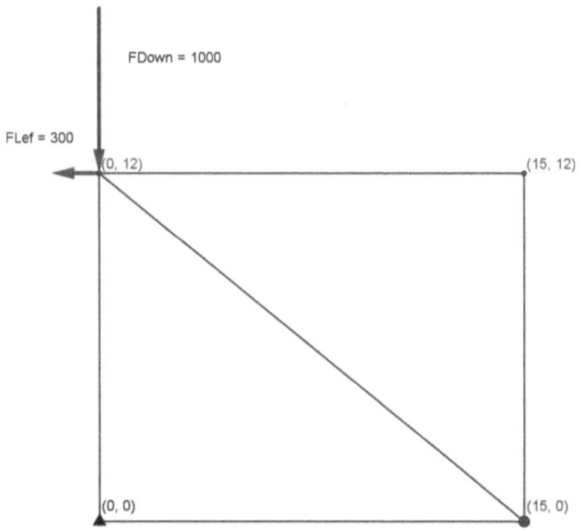

4.31
Étude 4.5 solution

Étude 4.6) For the truss in Figure 4.31, determine the left vertical reaction. A hand sketched solution is shown in Figure 4.32.

Étude 4.7) For the truss in Figure 4.31, determine the left vertical truss bar force. A hand sketched solution is shown in Figure 4.33.

Étude 4.8) For the truss in Figure 4.31, determine the diagonal truss bar force. A hand sketched solution is shown in Figure 4.34.

Étude 4.9) Three truss members, each are 7 units of length long, are brought to a construction site. The construction supervisor tells you to place the base of Bar1 at Cartesian Coordinates (0,0,0), the base of Bar2 at Cartesian Coordinates (8,0,0), and the base of Bar3 at Cartesian Coordinates (0,8,0). The construction supervisor then asks you to connect the other ends of Bar2 and Bar3 and to hoist up that triangle. Sketch out by hand the two instability mechanisms that remain. One is

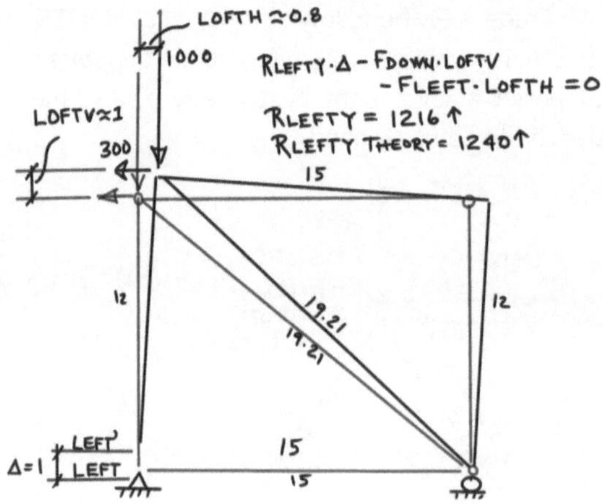

4.32
Étude 4.6 solution

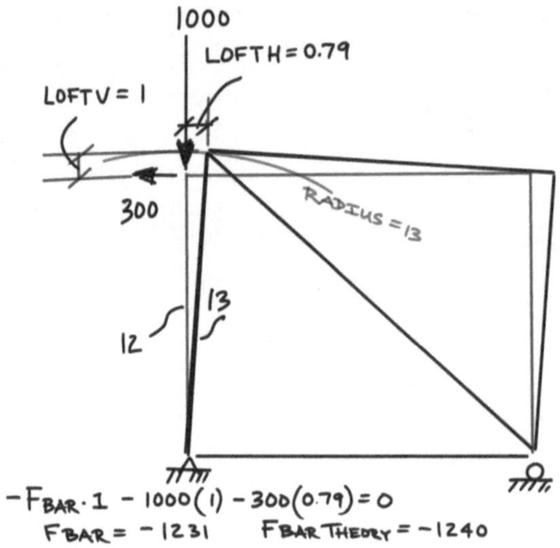

4.33
Étude 4.7 solution

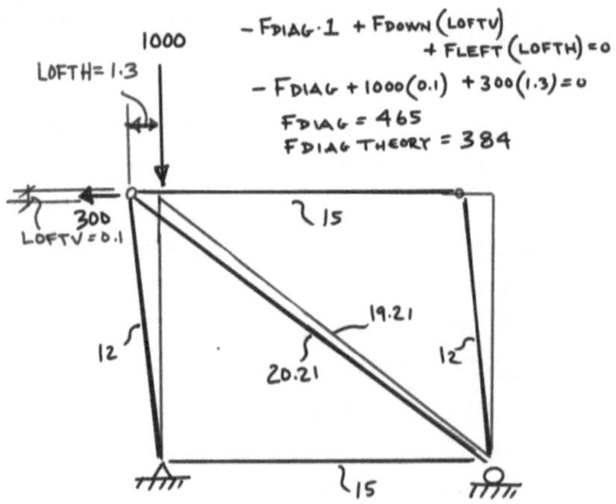

4.34
Étude 4.8 solution

4.35
Étude 4.9 solution
part 1

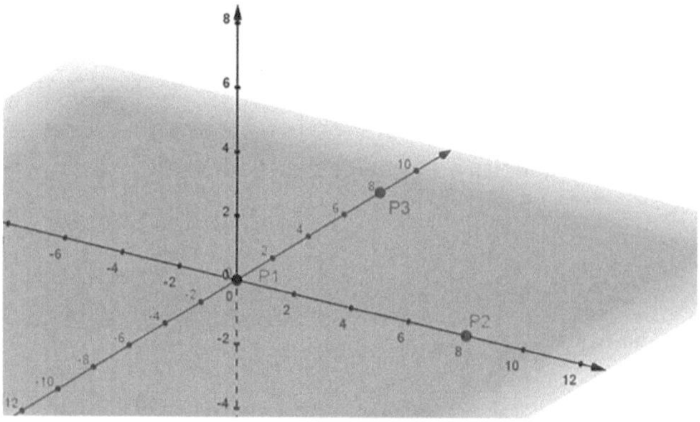

simply that of Bar1 by itself. The second is the mechanism for the triangle formed by Bar2 and Bar3 connected at some point above the ground. Then estimate as best you can the location of the final Crown meeting point. The base points are shown in Figure 4.35. A hand sketched solution is shown in Figure 4.36.

4.36
Étude 4.9 solution
part 2

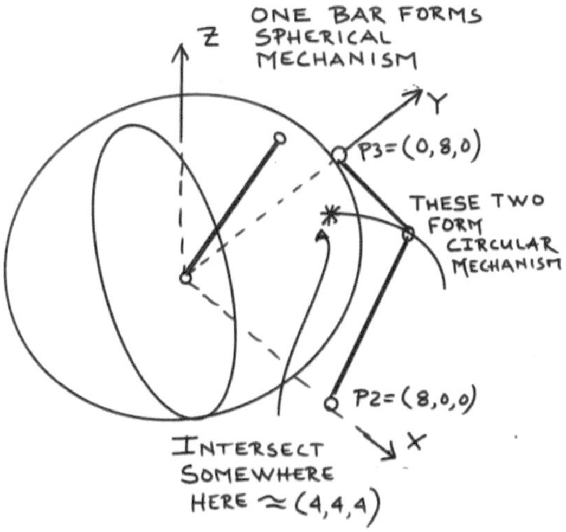

4.37
Étude 4.10 solution

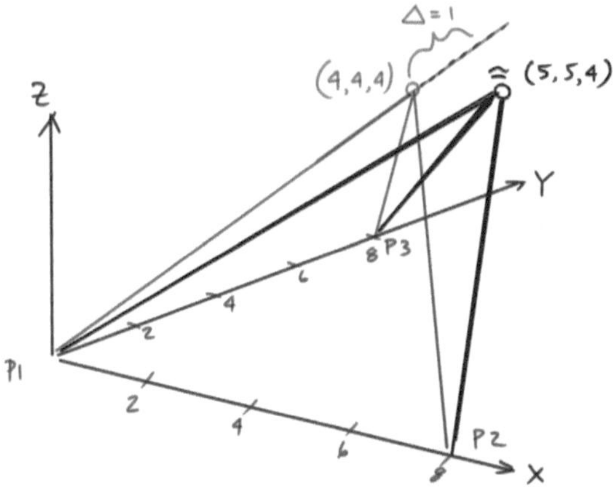

Étude 4.10) Start with the construction of étude 4.9, i.e. with three bars, each 7 units long, placed so that the bases are in the prescribed positions of étude 4.9, and the crown is approximately at (4,4,4). Elongate the Bar1 starting at P1 = (0,0,0) by an amount $\Delta = 1$ along its axis. Estimate with a hand sketch as best as you can the new position of the Final Crown. A hand sketched solution is shown in Figure 4.37.

5 GeoGebra for Trusses

To study the Modern Müller-Breslau Method as it applies to trusses, and to learn how to do so parametrically within a drawing program, consider the following truss shown in Figure 5.1. First, the reader should verify that the truss is determinate via Equation 4.1 of Chapter 4. There are nine bar elements, three equilibrating reactions, and six nodes. Note that the long diagonal crosses in front of, or behind, the shorter diagonals with no node at the crossing. Is it stable even though it is not strictly "triangulated" as is a regular truss? What typology of truss is this, according to the four categories presented in Chapter 5?

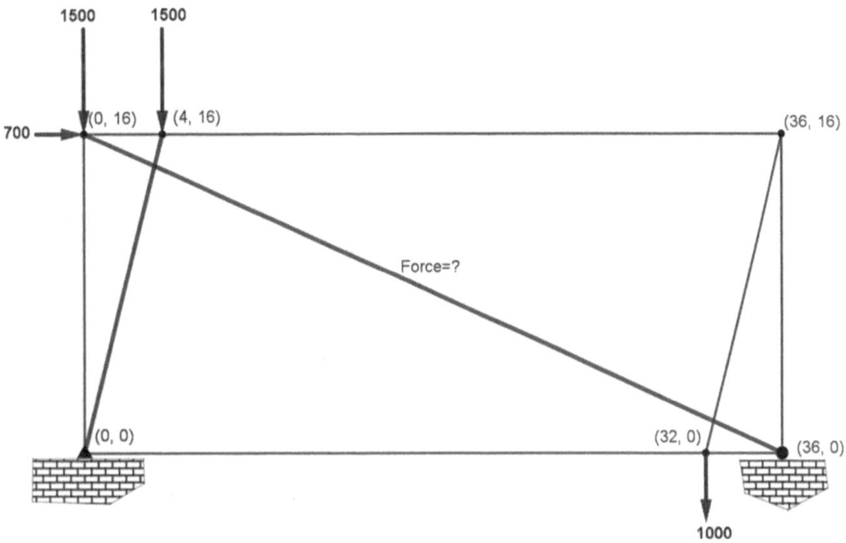

5.1
Compound Truss, find one bar force

We seek the internal bar force of the long diagonal, going from the top left to the bottom right nodes which originally is 39.4 units of length. As such, the Unknown is the axial force, Δ is the lengthening of the member being investigated. Figure 5.2 shows the distorted truss for $\Delta = 1$, which makes the new long diagonal 40.4 units of length. In a parametric drawing, Δ should be a variable that approaches zero.

DOI: 10.4324/9781003180913-5

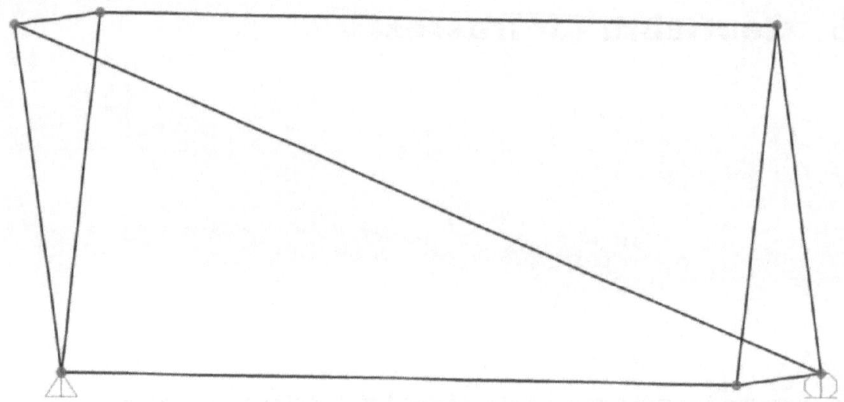

5.2
Perturbed Truss from a long diagonal stretch of $\Delta = 1$

One of the fundamental rules of determinate truss analysis is this:

- Only the bar whose unknown internal axial force is sought changes length, all other bars remain unstretched, or uncompressed; thus, the remaining bars have no axial force in them due to the perturbation.

This helps speed up the solution for any particular problem. The reader should always begin the process at a node that does not move, i.e. if there is a "pinned support", start there. Then, the reader should look to see if any "closed unambiguous loops" exist from that starting point. Figure 5.3 shows just such a loop, which would act as a rigid body. We know it is rigid and unambiguous because it starts and ends at a known point, and all the members in the loop have known, here unchanged, original lengths. Notice in Figure 5.3 that the rigid triangle formed by three members follows a circular path. In this rigid and perturbed triangle, the small top chord element connected to the leftmost member remains at 90° to the leftmost member. This is the only possible triangle that can be formed from the three original lengths of those first three elements.

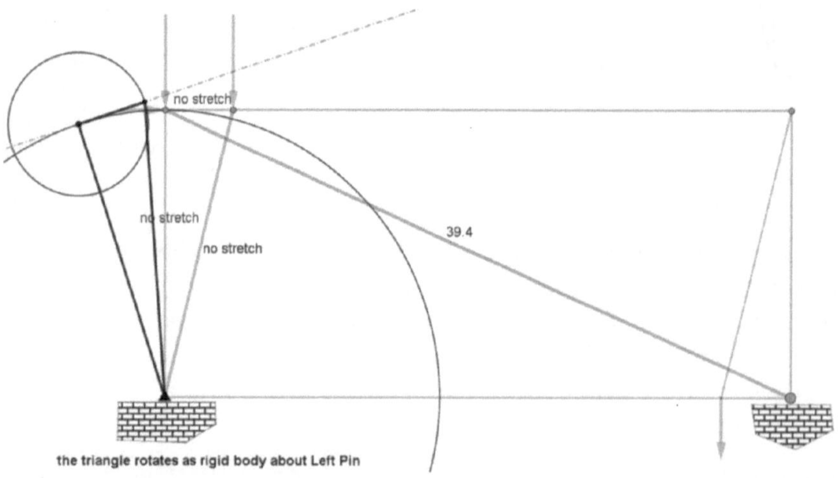

5.3
Start at a stationary point, find a closed loop

No other closed triangles exist branching off from any of these newly established, perturbed points. There is a quadrilateral that connects to the first rigid triangle, but that quadrilateral has two points with unknown locations; thus, it is not a closed unambiguous loop. But another closed triangle does exist at the end of the 32 unit top right bar. Thus, establish a linkage of 32 units along the top, then establish the second rigid triangle. This is shown in Figure 5.4. Here, several circles were used as opposed to a fixed angle.

5.4
Second closed loop exists

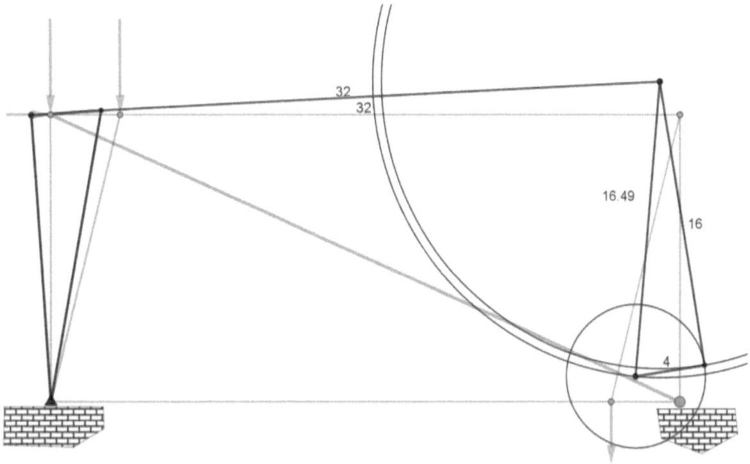

A helpful way of visualizing the closed loops is to think of them as rigid, solid triangles. Figure 5.5 demonstrates this for a very large perturbation, but notice that the perturbation has not been specified yet, it could be $\Delta = 1$ or it could be larger or smaller. In any script written for axial bar forces:

- Δ will be the difference in length between the original member being investigated and the axially perturbed member.

Figure 5.5 clearly displays two errors that have been created as we move around the truss. The first error is that the bottom bar should be 32 units of length. This error is induced by the circular path taken by the right rigid triangle. The second error is that the new right bottom node must be glued to the original horizontal surface. It is clearly floating above that surface in Figure 5.5. Both of these errors will be minimized in the subsequent and final step of the solution.

The structure is literally wiggled around till the error on the bottom bar is positive, but reasonably small and the error on the roller is positive but reasonably small. The way the problem was solved was by assuming positive errors. The signs in Equation 1.1 of Chapter 1 do indeed matter, thus the reader is encouraged to initially program assuming some sign for each loft and for the Δ. Figure 5.6 shows a reasonably large, yet still positive error for both the bottom bar and for the right roller.

After a few adjustments, the final answer is easily approached. This is shown in Figure 5.7.

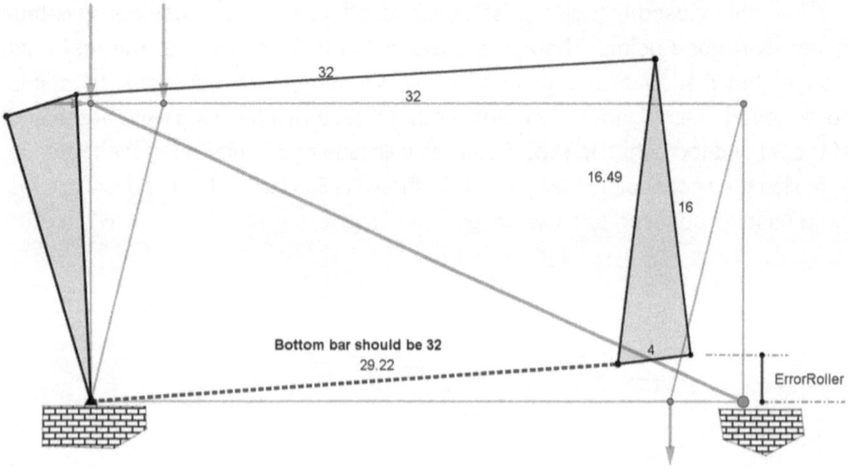

5.5
Two closed unambiguous loops exist

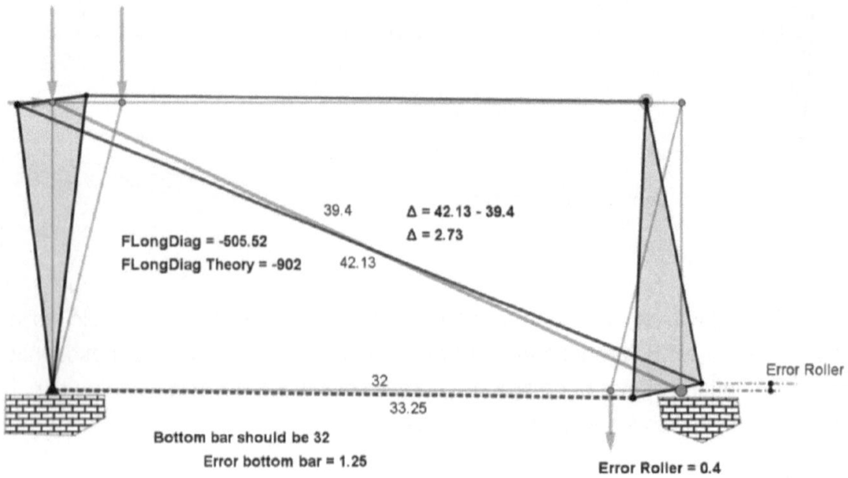

5.6
Final solution not yet reached

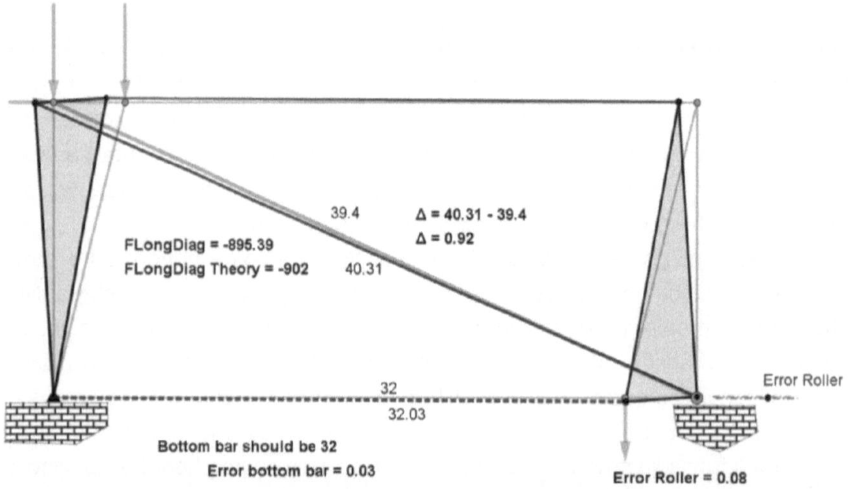

5.7
Final solution approached

The following drawing method is a rigorous way of establishing linkages. But it is time-consuming. A very fast way of solving these types of problems is to jump ahead to Figure 5.17 which uses the "rigid polygon" approach that was described in Figure 5.7. However, the reader who wants to get more familiar with a step-by-step approach to creating "linkages" is encouraged to study the following diagrams, which begins by considering the third truss of Figure 4.6 of Chapter 4. First, get comfortable drawing this truss in GeoGebra in its undeformed shape, but use parametric variables wherever possible, so that the drawing could be adjusted rapidly to study a similar truss. The original shape of the truss is shown in Figure 5.8 and the initial construction in GeoGebra is also shown. A new feature displayed in Figure 5.8 is the idea of so-called "grid lines". These are helpful and are used in construction drawings to identify key points in a structure. Grid lines will be used extensively in Chapter 6 where horizontal grids of beams are studied. In the initial setup of the parametric drawing, the spacing of the grid lines is clearly established. Then, after one point is laid down, subsequent points can build off of that first point, via the gridline spacing. Use a letter as the initial character of the variable, i.e. use "Space12" not "12Spacing" or similar names.

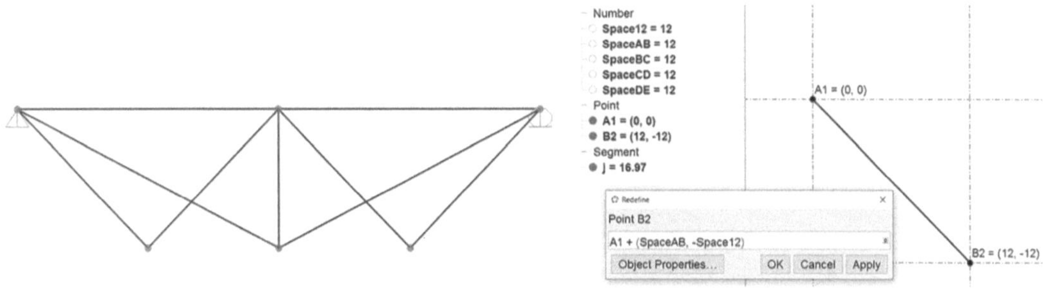

5.8
Setting up the drawing of a somewhat complicated irregular truss

The reader is now encouraged to simply try drawing in GeoGebra the original, undeformed shape of the truss on the left of Figure 5.8. As was described in Figure 5.1, in the truss of Figure 5.8 the crossing points are not nodes, and the bars pass either in front of or behind one another.

Figure 5.9 shows the original truss with three loads applied. Recall a rule of truss loads, namely:

- Trusses can only be loaded at nodes, never on the bars themselves.

Note that the single bar force between node A1 and node B2 is sought. The Cartesian Coordinates of all the nodes are displayed, as are the three equilibrating reactions, shown as icons at A1 and E1, in Figure 5.9.

To find the axial force in the bottom left chord member, here labeled as "bott1", perturb that member along its length by some amount Δ. The lower end of bott1 is not yet firmly established, but it must be on the circle which is centered on the Left support and whose radius is (bott1+Δ). This is shown in Figure 5.10.

All other boundary conditions must be respected. In a determinate truss, one truth which must be maintained in all problems is that the sole element whose

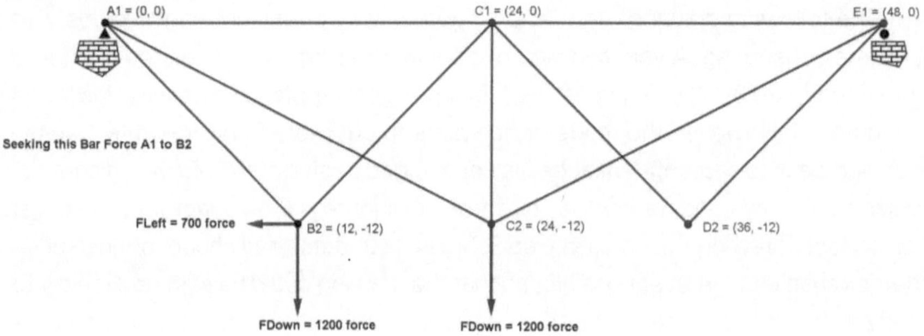

5.9
Setup for irregular truss with loads at some nodes

length changes is the perturbed element. All other bars of the truss must maintain their original length. This leads to an interesting insight for the perturbed condition of Figure 5.10. How can all the other elements maintain their original length?

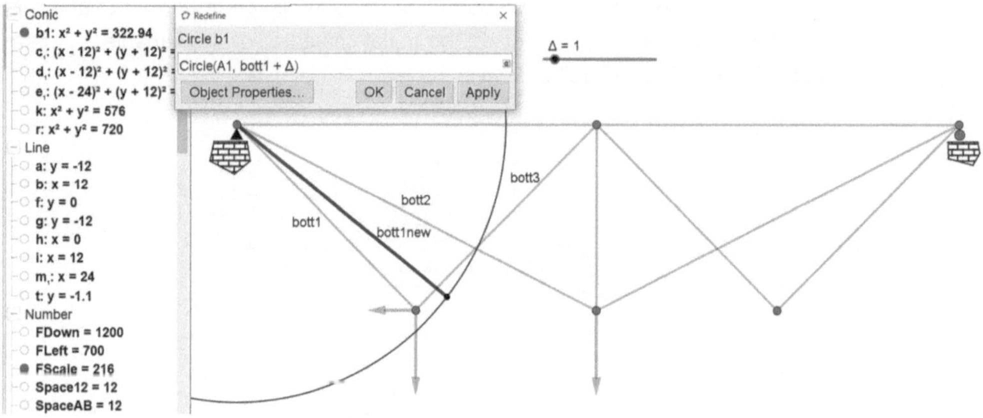

Consider building off of the end of the perturbed bar bott1new. One circle centered on the end of this bar must have a radius of bott3. Another circle, centered on the Left node, must have a radius of top1. Where those two circles intersect must be the top center node of the perturbed truss. This is shown in Figure 5.11.

5.10
Perturb bottom left chord by Δ

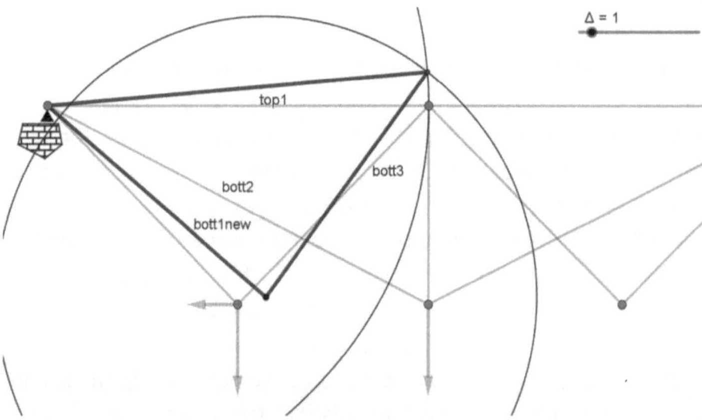

5.11
Find the next two bar locations

Construction of subsequent elements slowly proceeds. The intersection of circles ensures that the lengths do not change and that a linkage or "mechanism" is created which has great physical insights. Figure 5.12 shows the next two bars.

5.12
Two more bars located

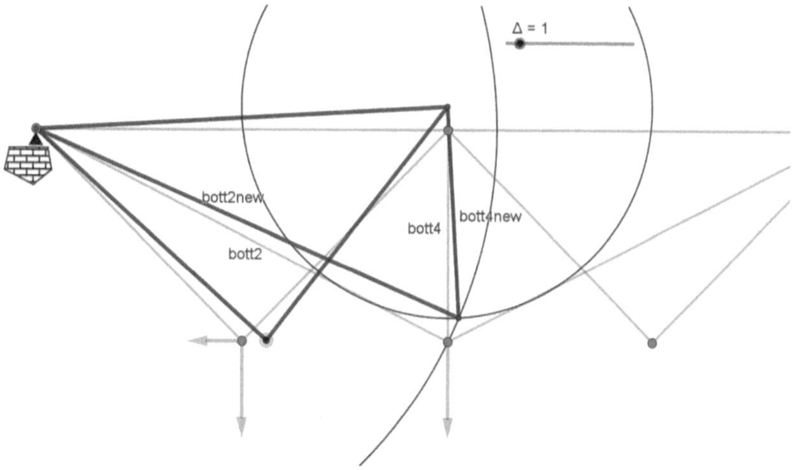

Figure 5.13 shows the final two bars as linkages. Notice that the entire truss moves as a rigid body. This is why we suggested that a faster way of drawing these perturbed trusses is to use rigid bodies. They will be explored shortly. Note in Figure 5.13 that the perturbed end of the rigid body does not rest on the original horizontal right surface, this must be corrected!

5.13
Final two bars located

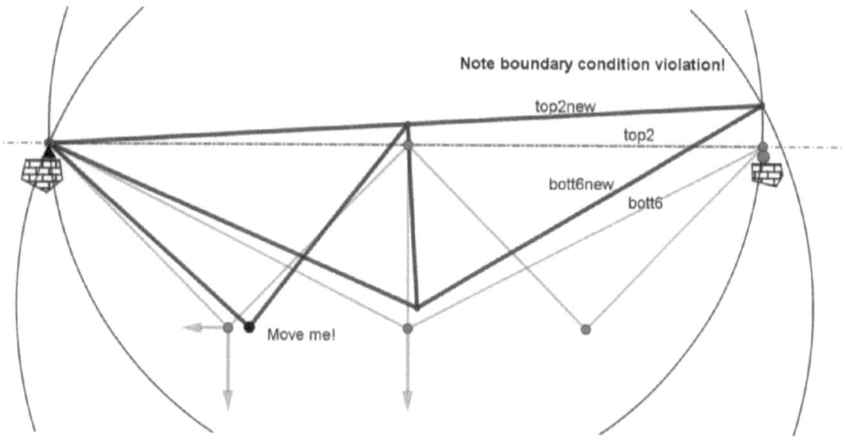

Figure 5.14 shows something that may be initially puzzling to the reader. The boundary condition error on the right roller is clear and apparent, as the rightmost node must be glued to the original horizontal supporting surface. But since the entire truss moves as a rigid body, as was established in Figure 5.13, the reduction of that boundary condition error can be achieved simply by moving or wiggling the entire rigid body. Yet, this reduction of the boundary condition error has nothing to

do with the perturbation Δ. The perturbation Δ is along the unknown bar force's axial length, and in Figure 5.14 it is still large.

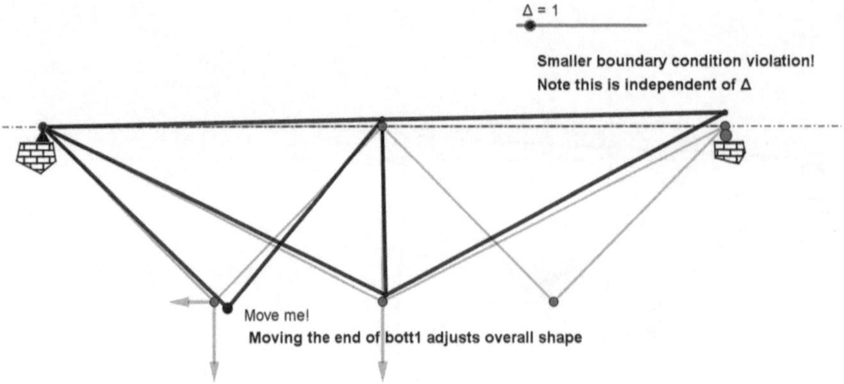

5.14
Reducing the right boundary condition error

The error is reduced and the problem is solved via Equation 1.1 of Chapter 1, as modified by Equation 4.3 of Chapter 4. It becomes immediately apparent that the centrally applied vertical downward force has absolutely no effect on the bar force of the bottom left chord member, bott1. Further inspection of the circuitous load path that the central vertical downward force might provide some insight, yet it is challenging to see this lack of influence via traditional methods. In Figure 5.15, the Modern Müller-Breslau Method has immediate visual verification that the central load does not loft at all due to a perturbation of the bottom left chord member, bott1. Had the reader noticed this lack of deformation in the other members, the simplification of using rigid triangles, as was done in Figure 5.5, would have sped the process up.

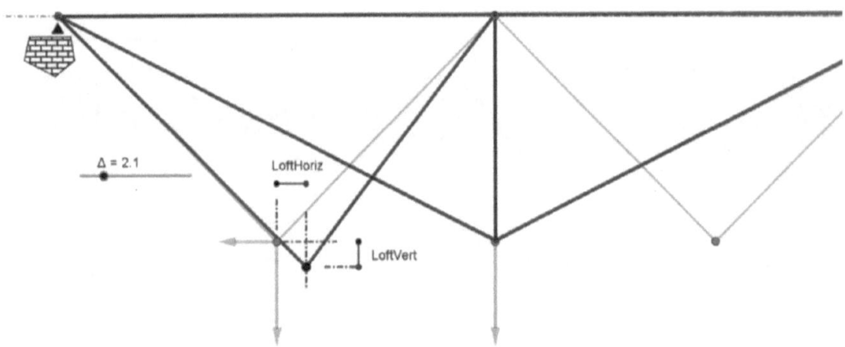

5.15
Notice that only one bar deforms

While the drawing of segments "LoftHoriz" and "LoftVert" in Figure 5.15 are visually instructive, it may be easier for the reader to simply measure lofts, or even the error of the roller placement, as simply the difference between the Cartesian values of two points. This is demonstrated in Figure 5.16, where an error is explicitly measured as the difference between the y values of two points.

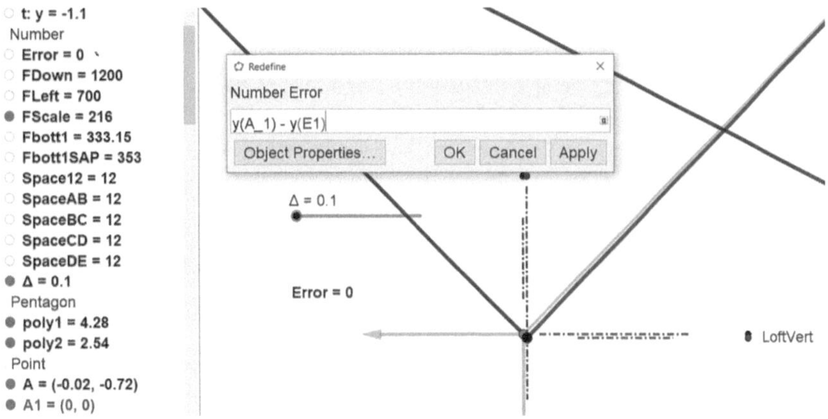

5.16
Distances can be measured as segments or as the difference of *x* or *y* values of two points

The fact that only one bar elongates during the study of any determinate truss using the Modern Müller-Breslau Method makes the construction of the distorted shape reasonably quick with rigid closed loops of elements, which are best thought of as rigid polygons. Consider, for example, Figure 4.9 of Chapter 4. It is redrawn as Figure 5.17 here, with the rigid polygon highlighted. A useful tool in GeoGeobra is the ability to create rigid polygons. The easiest way to draw the images of Figure 5.17 is to input the Points, always starting a Point name with a capital letter, and the Points can be input with simple Cartesian Coordinates, or with parametric gridline line spacing, for example, *B*2 = (15, 7.5) or *B*2 = (spacingAB, spacing12) where spacingAB was previously defined as 15 and spacing12 was previously defined as 7.5.

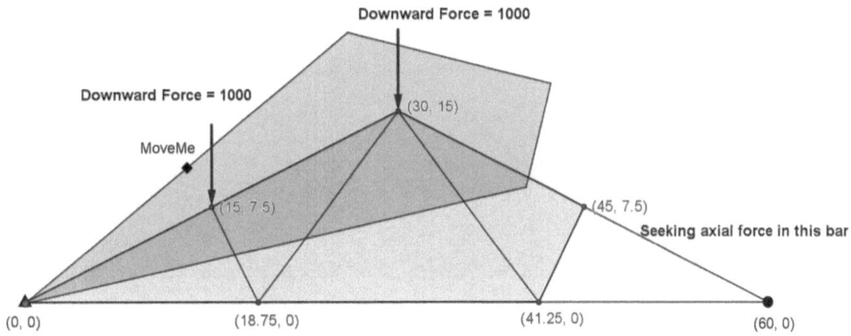

5.17
Rigid polygons speed up the drawing process

The fastest way of creating such a rigid polygon, is to first draw a polygon by clicking on the first node which will be the pivot point, here clearly that is the lower left pin support, and then clicking subsequent nodes until you click again on the starting point, which will be the pivot point of the rigid polygon. GeoGebra has a handy icon, it is the third dropdown under the Polygon Icon Menu on top, and it is called Rigid Polygon. It defaults to triangles, but if one has pre-defined a polygon, then simply clicking on that pre-defined polygon with the Rigid Polygon menu active, will create a copy of the original polygon. The new Rigid Polygon will pivot about the first point. The second point, which can be conveniently renamed as MoveMe, will allow for the spinning of the Rigid Polygon. That polygon can be repositioned such that its pivot point is very close to pinned support. The copied polygon is shown in Figure 5.18.

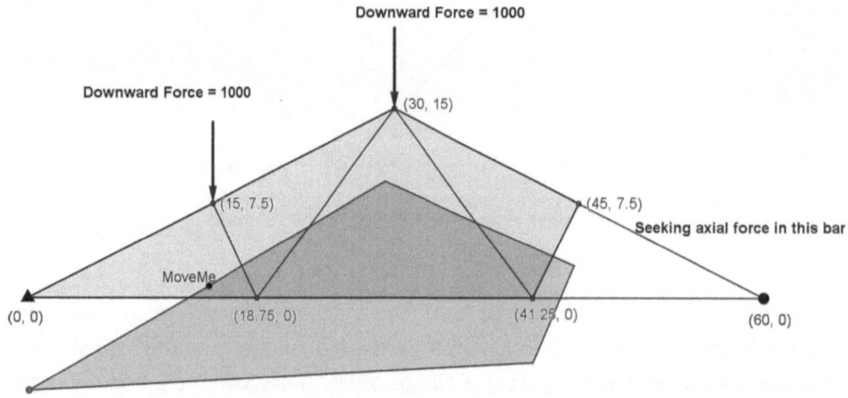

5.18
Rigid polygons may be used to speed up the drawing process

This chapter closes with some qualitative bar force geometric solutions for a few trusses that were briefly described in Chapter 4. Figure 5.19 shows one such truss, with combined gravity and lateral loading. The bar force on the lower left chord is sought, as is the bar force in one of the central web members. The Modern Müller-Breslau Method can find both of these, but only one at a time.

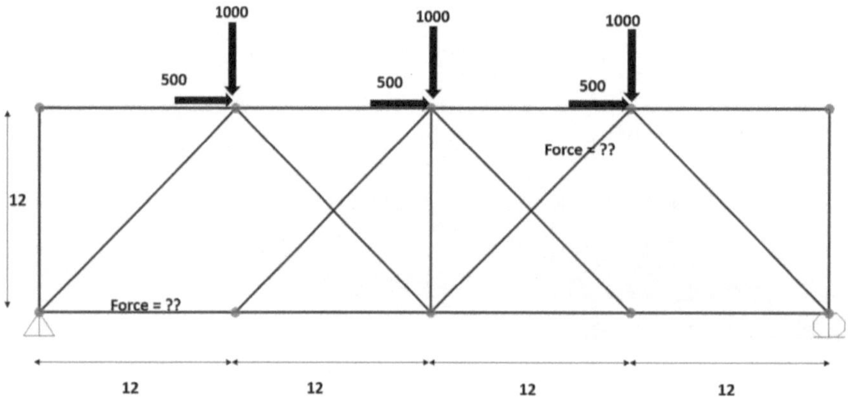

5.19
Gravity and lateral loads on the truss

The geometric solution for the lower bar perturbation is shown in Figure 5.20. The reader is encouraged to study this solution, sketch it by hand, and then see what rigid polygons exist in the perturbed structure.

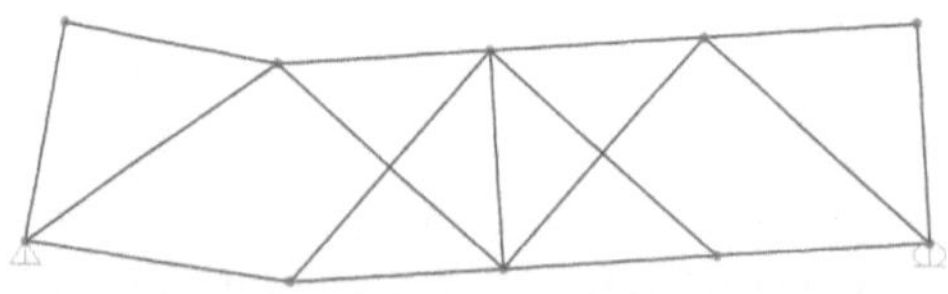

5.20
Qualitative solution for lower left chord stretch Δ

Clearly, two rigid polygons are present in Figure 5.20. The triangle on the left side, and the large, nearly rectangular shape on the right side. Separating these two rigid polygons is the perturbed lower left chord member whose axial force is sought. The lower left chord member can be compressed an amount Δ as in Figure 5.21, or elongated some amount Δ, as in Figure 5.22. Note, the rigid polygon method easily can accommodate a Δ that is compressed, or elongated, but in this textbook, for consistency Δ is always an elongation. That way, if the answer is positive, the force is tensile.

5.21
Rigid polygon solution for lower left chord compression Δ

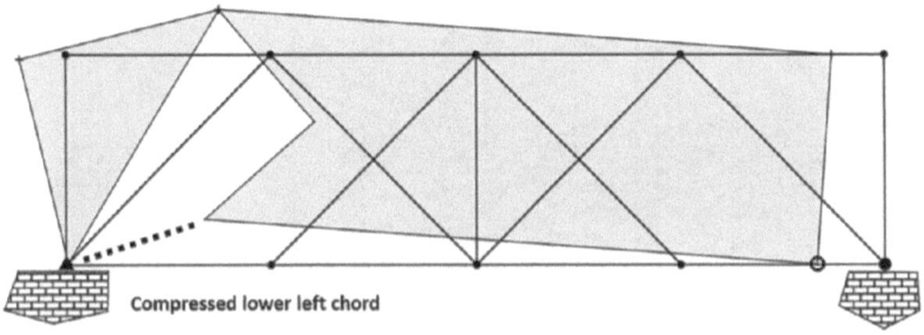

5.22
Rigid polygon solution for lower left chord elongation Δ

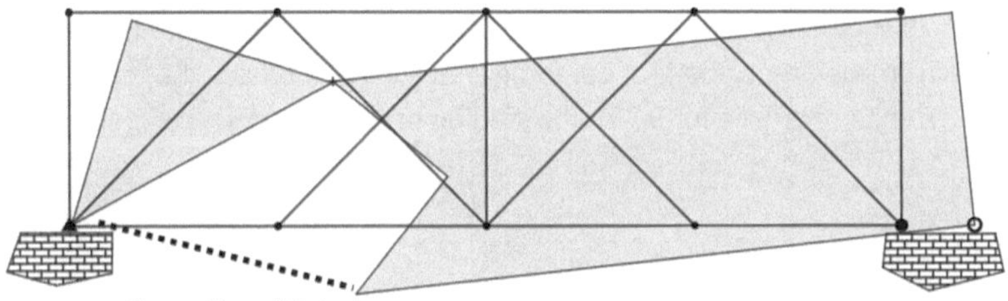

In all such perturbations, large sections of the truss move as rigid polygons. This truth lends itself to rapid physical modeling and geometric solutions. The technique can be applied to find the force in any member of the truss. Figure 5.23 shows a rigid polygon solution for an interior web member's force.

5.23
Solution for interior web member stretch Δ

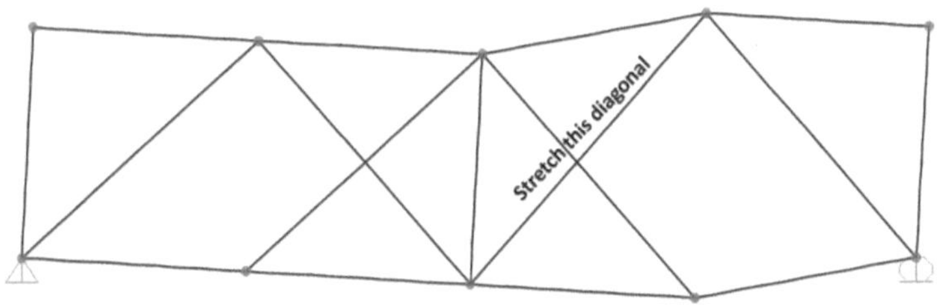

Rigid polygons exist, but since there are three distinct members connecting the left and right polygons, and only one of them stretches (the one whose force is being sought), the other two members have unambiguous, unchanging lengths. But notice that the two known length bars, both of which are 12 units of length long, have ambiguous and unknown orientations. Yet, a fairly straightforward way of ensuring they both remain at 12 units of length is to draw two circles centered on the left polygon's connecting points. The right-hand polygon must then touch each of these circles. This is shown in Figure 5.24.

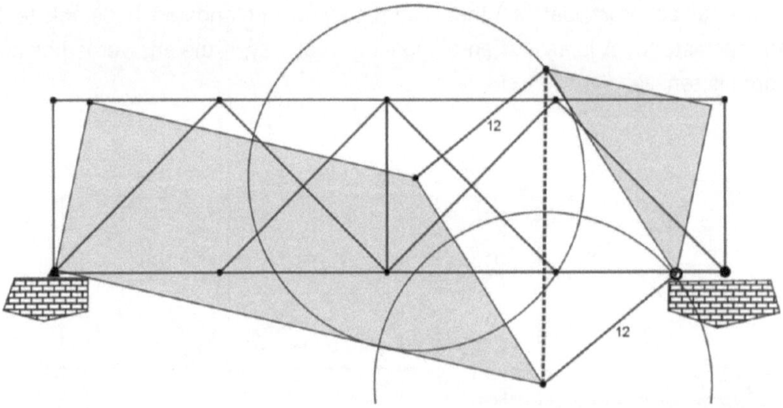

5.24
One way of connecting the two rigid polygons

Figure 5.25 shows the same construction as in Figure 5.24, but here the perturbation is much smaller; thus, the final bar force answer will approach the theoretical value of 1,237 units of tensile force, for the loading originally shown in Figure 5.19.

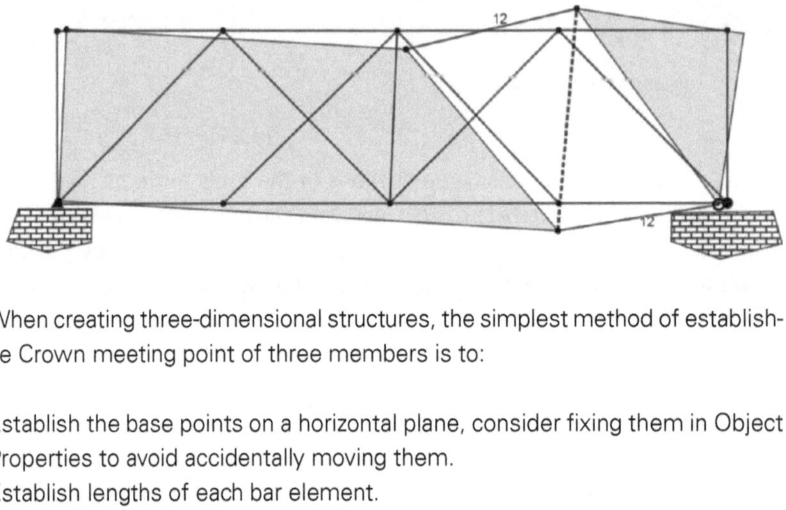

5.25
Approaching the final answer

When creating three-dimensional structures, the simplest method of establishing the Crown meeting point of three members is to:

- Establish the base points on a horizontal plane, consider fixing them in Object Properties to avoid accidentally moving them.
- Establish lengths of each bar element.
- Establish a sphere at each point, whose radius is the associated bar length.
- Find a circle, that is the intersection of two spheres, for example Circ12 = Intersect(sphere1,sphere2) would be a good line of code to write if sphere1 and sphere2 have been defined by the previous step.
- Find a Point that is the intersection of that circle and the third sphere, for example, Crown = Intersect(Circ12,sphere3). Notice that one intersection is above ground, and one is below ground. Hide, do not delete, the Crown that is below the horizontal ground plane.

Figure 5.26 shows an elementary tripod in 3D space. Points P1, P2, and P3 have been defined and fixed. Spheres of radius 4 and 7 and 5 are centered on Points P1, P2, and P3, respectively. A circle which is the intersection of two spheres is created, then hidden. The third sphere intersects with that circle to create the Crown above ground. The Crown below ground is hidden.

5.26
Setting up an elementary tripod

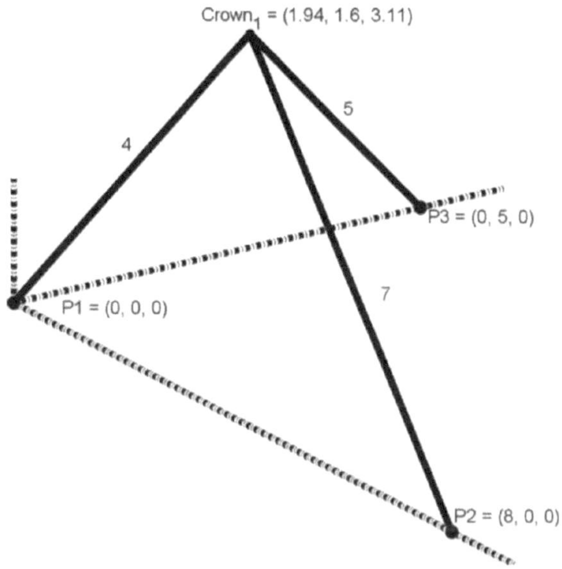

To find a vertical or a horizontal equilibrating reaction at P1, P2 or P3, first note that each of these Points has three unknown possible reaction forces, depending on how the Crown is loaded. Any single reaction can be found by perturbing that single reaction an amount Δ. For example, suppose we sought the vertical reaction at P2 in Figure 5.26. To find this reaction, we would perturb P2 vertically, but not horizontally, creating P2′ which is some Δ above the original horizontal plane. Essentially the previous steps would be recreated, with the original configuration hidden, or grayed out. The new tripod would be found exactly as the original tripod was, but now P1, P2′, and P3 are the base points. Figure 5.27 shows the start of a graphical solution for this perturbation, with some loads applied at the Crown. Notice that P2′ is directly above P2, perturbed by a known amount Δ.

Using the same steps as before, the new sphere around P2 of radius 7 is used to find NewCrown. A fast way of finding the lofts is to simply capture the difference in Cartesian Coordinates between the NewCrown and the Crown. But be careful with the signs! Assume positive is aligned with axes, then the force in the *Z* direction would be – 800 in Figure 5.27. The quickly captured lofts are described in Figure 5.28.

Finally, the vertical equilibrating reaction at P2 is quickly found. The solution is shown in Figure 5.29 for a reasonably small $\Delta = 0.5$. As Δ gets smaller, the final bar force approaches the theoretical value.

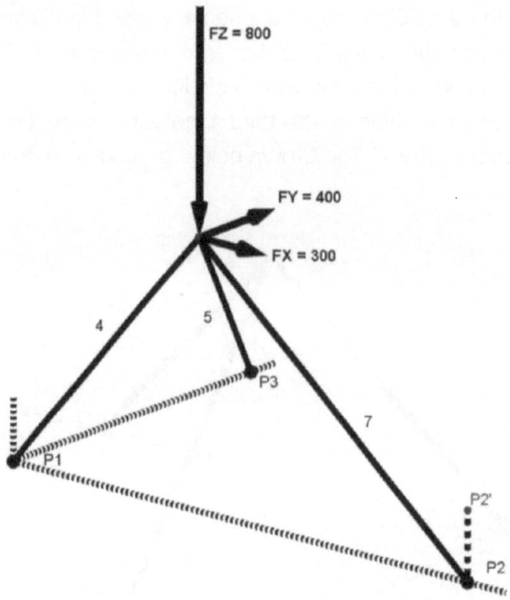

5.27
Loads applied to a tripod

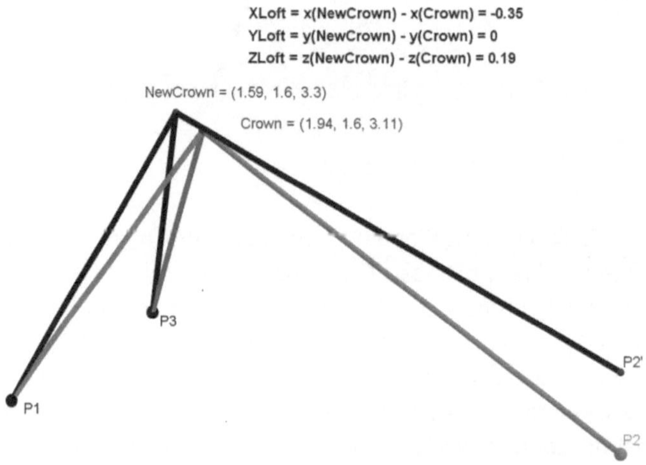

5.28
Equilibrating vertical force at *P*2, large Δ

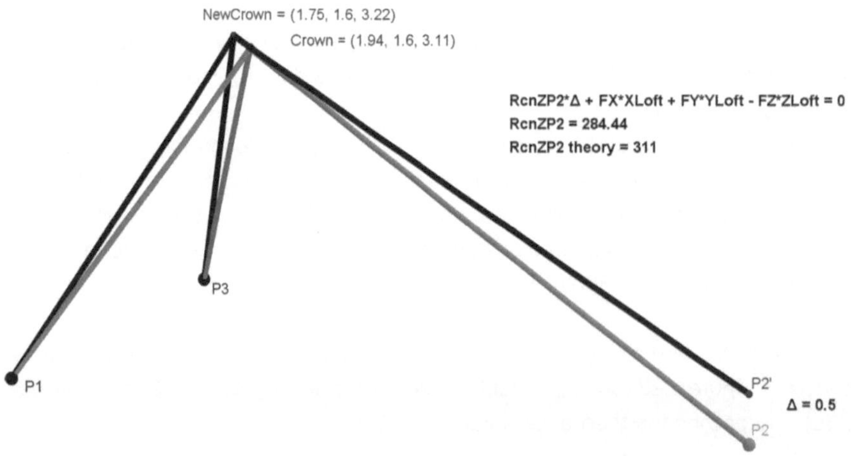

5.29
Equilibrating vertical force at *P*2, small Δ

Finding any particular equilibrating internal bar force is quick to do, once the initial drawing has been set up. If using a previous script, consider creating a new variable Δaxial to introduce a perturbation to a bar, and not to an external reaction. Create a sphere of radius (OriginalLength + Δaxial) around the base point of the element studied. Then proceed as before with an intersection of two spheres to get a circle, and the intersection of the circle and the third sphere to get the perturbed Crown. As Δaxial is reduced, the final bar force approaches the theoretical value.

5.30
Axial Force in Bar 2, reasonable Δ

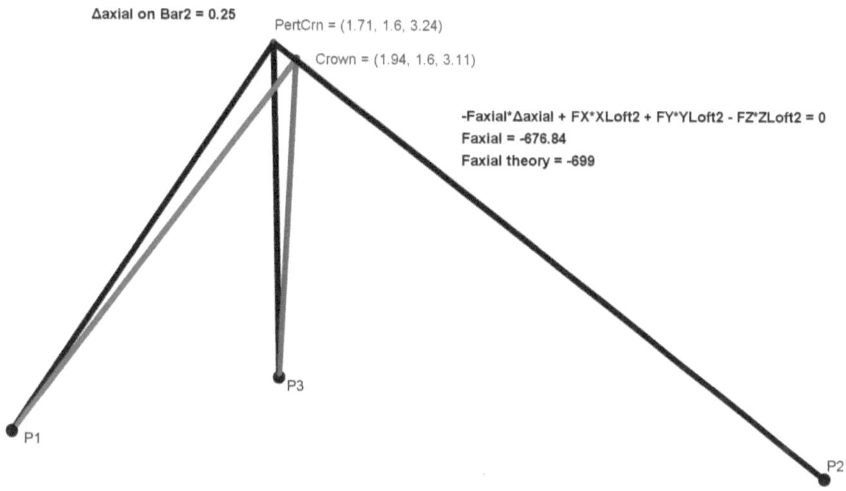

CHAPTER 5 ÉTUDES

The following studies or études should be done in GeoGebra.

Étude 5.1) This is a revisiting of étude 4.1, with a few extra steps. A truss has three members, the top chords are each 10 units long and the bottom chord is 12 units of length. The truss is pinned at the left end and is roller supported vertically at the right end on a single horizontal plane between Left and Right. There is a downward load of 500 and a leftward load of 250 units of force applied to the crown. Find the Cartesian Coordinates of the Crown.

Part 1) Input all parameters such that your script can be rapidly reconfigured for other similar structures. The only entity that should be fixed is Left = (0,0). You can even use Object Properties to Fix Object to ensure you don't accidentally move Left. Part 1 is to simply establish the geometry including the unique Crown location. Figure 5.31 shows the nearly completed geometry. Other solutions that the readers create are most welcome, just keep everything parametric.

Étude 5.2) Repeat étude 5.1, but this time use the parametric script to immediately find a new Crown if the bottom chord member is 14 units long, rather than 12. Your script should instantly update once that parameter is changed to 14. Figure 5.32 shows the completed geometry.

Étude 5.3) Repeat étude 5.1, but also add a rigid polygon, which is a triangle in this situation. Perturb the rigid triangle so that it pivots about the left pinned support and this creates a vertical Δ of the right support, which is captured from

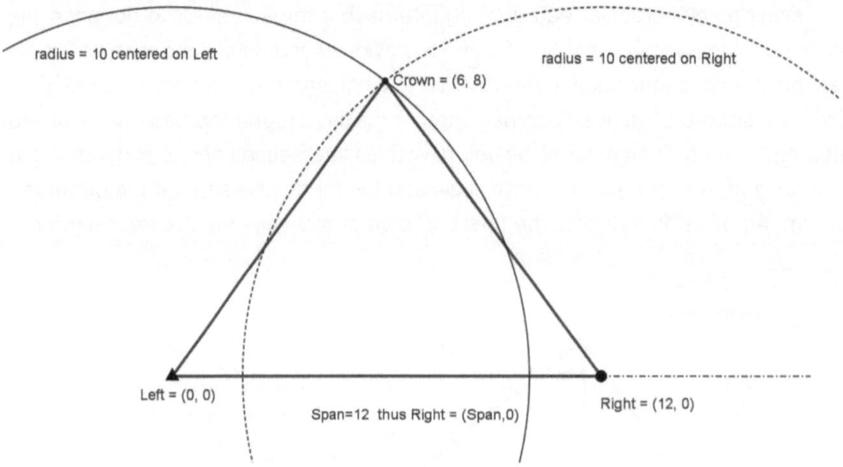

5.31
Partial solution to étude 5.1

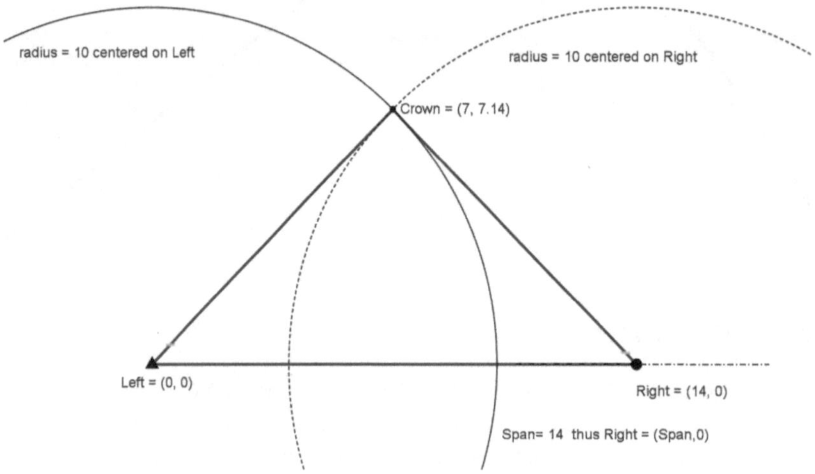

5.32
Solution to étude 5.2

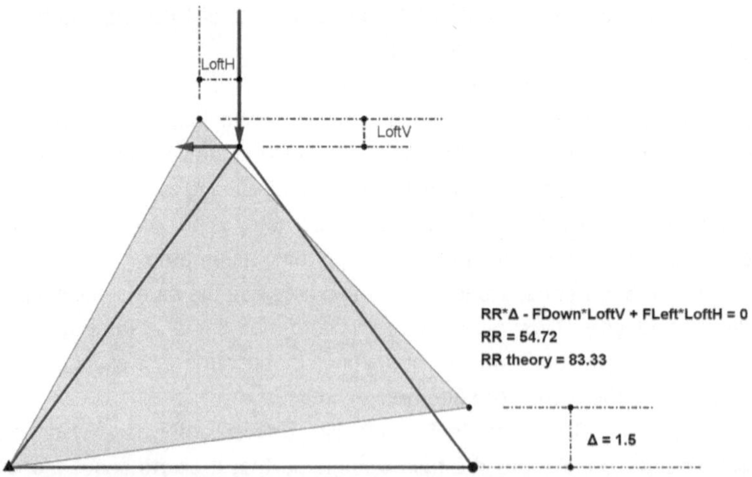

5.33
Solution to étude 5.3, large Δ

the circular path that the right support follows. Use all this information to establish the magnitude of the equilibrating vertical right reaction, and do so for a fairly large Δ, then for a small Δ. A solution for large Δ is shown in Figure 5.33. A solution for small Δ is shown in Figure 5.34.

5.34
Solution to étude 5.3, small Δ

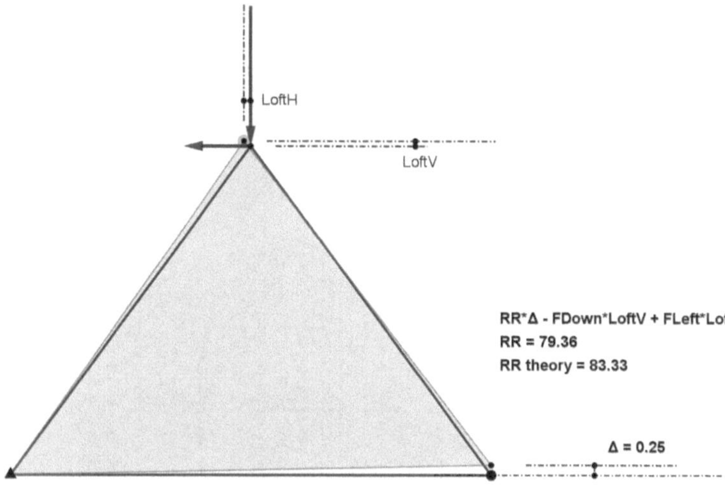

Étude 5.4) Repeat étude 5.1, but now stretch the top left chord a variable amount Δ. Establish the perturbed shape. Unless you have a more efficient way of doing this, type in, via the Algebra window or via a slider a number Δ. Calculate the new Top Left Chord Length as TopChLength = 10 + Δ. Then, draw a circle of radius TopChLength centered on the Left. Drop a point on that circle, call it NewCrown. Draw a circle centered on NewCrown radius 10. Draw a circle centered on Left radius 12. The intersection of these two circles is the new Right'. Adjust the diagram by moving NewCrown till Right' has a very small, but positive error, i.e. till it is barely above the horizontal $y = 0$. Figure 5.35 shows a solution for a Δ = 1. Recall that Δ is the distance between NewCrown and Crown.

5.35
Solution to étude 5.4

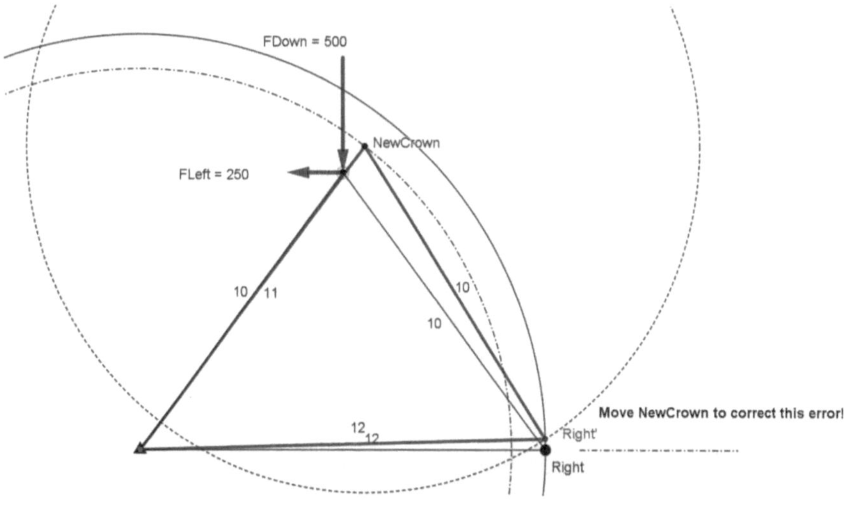

Étude 5.5) Repeat étude 5.1 as well as étude 5.4 to quantitatively establish the internal axial force in the top left chord member that was stretched in Figure 5.35. A solution is shown in Figure 5.36.

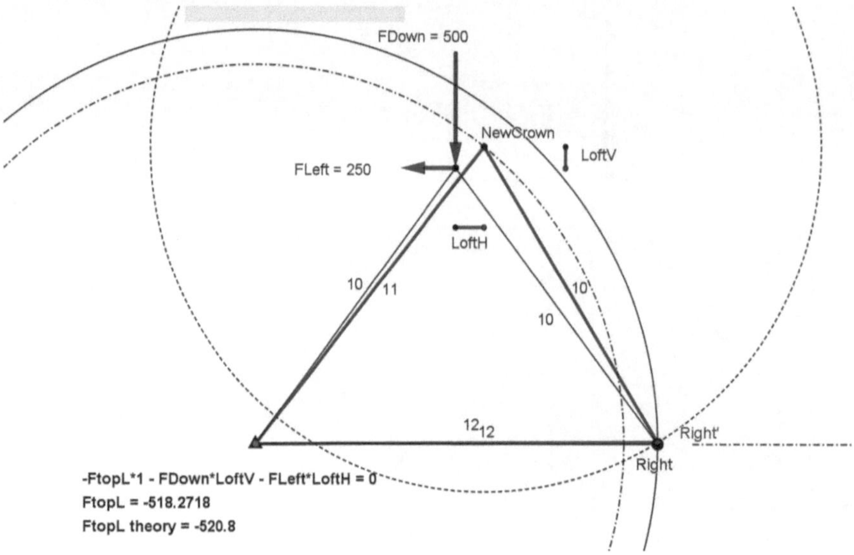

5.36
Solution to étude 5.5

Étude 5.6) A tripod made up of three pin-ended members, each with 9 units of length long is set up in 3D. The Cartesian Coordinates of the base points are $P1 = (0,0,0)$ $P2 = (6,0,0)$ $P3 = (0,6,0)$. Find the Crown meeting point. A solution is shown in Figure 5.37.

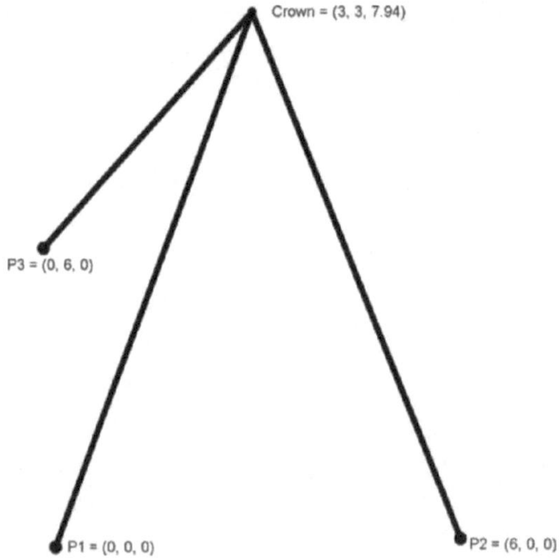

5.37
Solution to étude 5.6

Étude 5.7) Apply a single downward load Fz = 1,000 at the crown of the tripod in Figure 5.37. Find the vertical equilibrating reaction force at P2. A solution is shown in Figure 5.38.

5.38
Solution to
étude 5.7

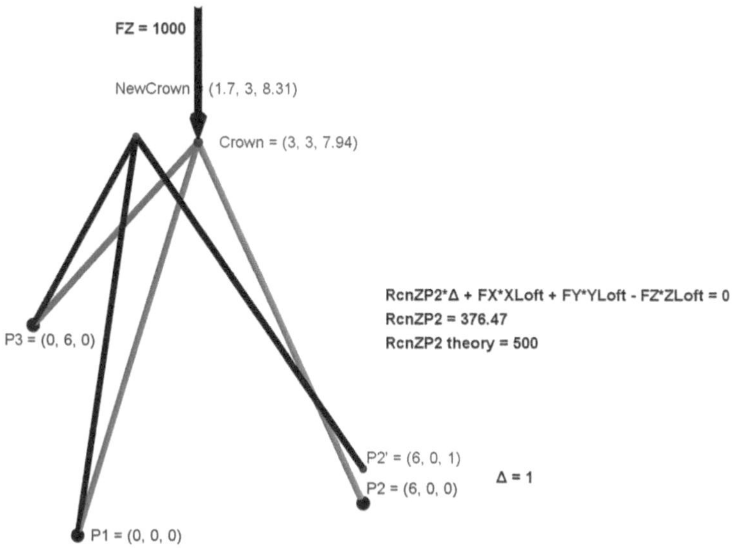

Étude 5.8) If the theoretical vertical reaction at *P2* and at *P3* are both 500 units of force upward in Figure 5.38, what must be the theoretical vertical equilibrating reaction at *P1*? Explain if this seems reasonable or not.

Étude 5.9) For the original tripod and single downward load shown in Figure 5.38, find the equilibrating axial force in Bar2, whose base is *P2*. A solution is shown in Figure 5.39.

5.39
Solution to
étude 5.8

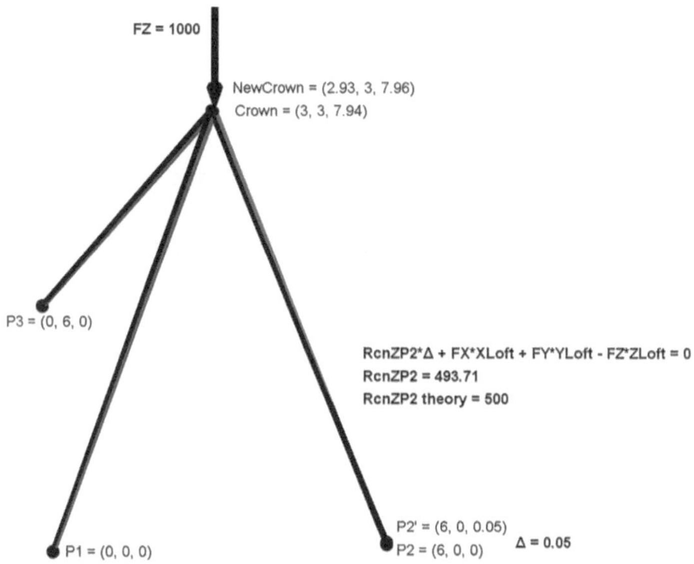

Étude 5.10) Given your answer for étude 5.7, what must the equilibrating axial force in Bar1 be for the tripod and load shown in Figure 5.38?

6 Horizontal Grids

A horizontal grid of beams is a very important structural configuration. It is very commonly used, and it will be completely determinate if the ends of individual beams do not transfer moment to the subsequent supporting member. Timber construction typically assumes such connections, i.e. simply supported ends. Occassionally, a cantilever will exist and a cantilever example will be studied in this chapter. But the most common case is a grid, or a horizontal network, of simply supported beams. A structural hierarchy within this network or grid must be understood, as it is critical to the topic of "load flow". Load flow is the investigation of how load travels from a floor or a roof that is supported by a horizontal grid of beams, all the way down to the columns, which subsequently pass the load down to the foundations. Notice that the load flow begins with the smallest members known as "joists". These are simply connected, without moment transfer, to other elements known as "beams". The beams can then connect simply to columns, or perhaps they connect simply to ever more important members on the hierarchy, known as "girders". This lingo of "joists", "beams", and "girders" is part of the practice of structural engineering, but if the reader finds it helpful to think of them all as beams, then the following hierarchy still must be maintained:

- The lowest level in the hierarchy is the joists or the smallest beams, and they support the roof or the floor.
- The next level up in the hierarchy is beams. Joists connect without moment transfer to beams.
- The next level up in the hierarchy is girders also known as "heavy beams". Beams connect without moment transfer to girders. Alternatively, some beams can connect without moment transfer to columns.

The highest hierarchical member in the grid is the girder. It connects to columns, without moment transfer. The girder carries the beams, which carry the joists, which carry the distributed floor or roof load.

Grid lines are essential in structural drawings. The intersection of grid lines denotes important structural elements, such as columns. Grid lines need not be evenly spaced, they need not be north/south and east/west. They may even be radial as the situation demands. Grid lines must inform the builder precisely where the important elements are. Figure 6.1 shows a typical horizontal grid of beams that is determinate. Notice the grid lines and notice the gaps at the ends of elements.

DOI: 10.4324/9781003180913-6

The gap is a universally accepted drawing convention that signifies the connection as being "simple" or in other words, the connection does not transfer moment to the subsequent member. The gap states that this connection is essentially a "hinge".

In Figure 6.1, the shortcut DL stands for Dead Load, i.e. load that is permanent and unwavering. It is expressed as force/length^2 or force/area. LL stands for Live Load, i.e. load that comes and goes, it is not permanent. Structural engineers handle DL and LL differently in that LL can be reduced by a standard procedure, which is based on the probability of all the LL not being fully applied at all times. LL reduction will be ignored in these examples to avoid intimidating the novice reader. But accomplished practitioners who study these examples may well marvel at the simplicity of obtaining tributary areas, which are necessary for applying LL reduction. As a final note on LL reduction, be aware that reducing the roof Live Load is handled slightly differently than reducing the floor Live Load. That distinction is not important in the following examples.

6.1
Horizontal grid of beams setup

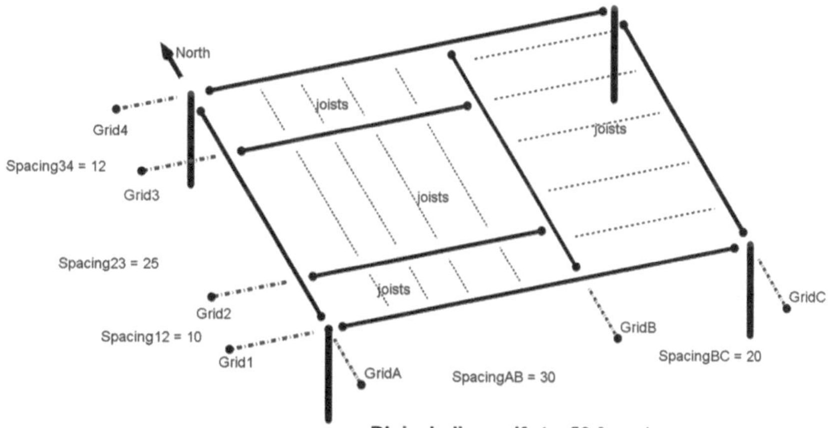

The reader should take time to notice several details. The first detail to notice is that joists run along one direction between grid lines A and B, and then they run along a different direction between grid lines B and C. Next, notice that the joists do not transfer moment to the beams, that is signified by the gap at the ends of the joists. Individual joists are not labeled, nor is the spacing of the joists important in this example. Joists have the lowest hierarchical importance. The load will flow much like water, imagine the floor (not shown) as being the source of the water poured downward. That water fills "pipes" that are joists. Those joists then pass the water into the next higher element in the hierarchy, namely the beams. The reader need not quibble about distinctions between "beams" and "girders", but the beams on grid lines 2 and 3 are good examples of this hierarchy. Joists connect to these beams from either side, passing load to these beams. Yet, what supports the beams on grid lines 2 and 3? Not the columns, but elements higher up in the hierarchy. The element along grid line A might be labeled as a girder, as it supports one end of the beams on lines 2 and 3. The element along grid line B could be labeled as a girder, it supports the other end of the beams on lines 2 and 3. Or some might label the elements on grid line A and on grid line B as beams, with the special name of "edge beam" applied to the beam on grid line A. Yet the names are not important here, the hierarchy is vitally important. The elements on grid lines A and B clearly support the elements on grid lines 2 and 3. That is what is meant by hierarchy. Note in Figure 6.1 that all the beams are simply supported, either by subsequent girders or by the columns, all with hinged, non-moment transferring connections. Finally, note that north is parallel to the lettered grid lines.

Figure 6.2 shows a somewhat simpler arrangement of beams in a horizontal grid. Here, there are girders or very large beams spanning along grid lines 1 and 2, these are primary. They support secondary beams, spanning along grid lines A, B, C, and D. These secondary beams support tertiary members which might be called joists, or might be a flat member like a floor or roof, and the load from these tertiary joists, which are parallel to grid lines 1 and 2 flow into the secondary beams.

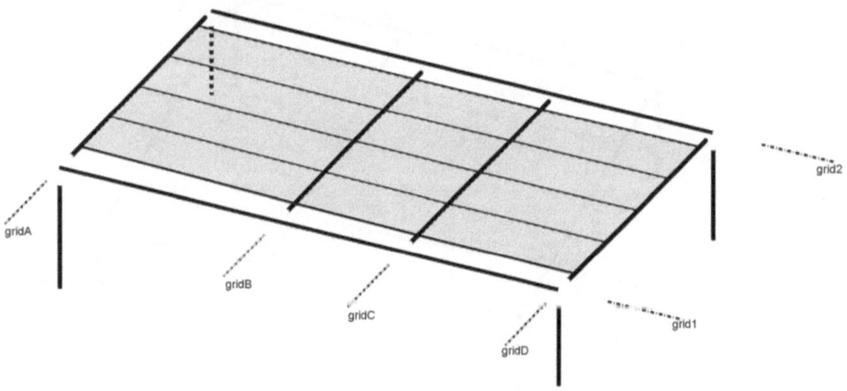

6.2
Typical horizontal grid supporting a floor

It may or may not be intuitive to the reader that ¼ of any total downward load uniformly applied to the grid in Figure 6.2 would flow to each column. Much like a balanced table, the symmetry makes it fairly intuitive to understand this concept. But how much load flows to any particular beam, for example, the beam on grid line B? For experienced readers, the terms "tributary width" and "tributary area" may come to mind. For the novice reader, the term "tributary area" is a concise way of assigning the load to any particular element as arising from a specifically designated area, i.e. the "tributary area". Think of the tributary area as a zone of influence, which flows into the element being studied. Thus, for any column in Figure 6.2, the tributary area would be ¼ of the total area. But what is the tributary area of the beam on grid line B in Figure 6.2? The Müller-Breslau Method answers this visually as shown in Figure 6.3. As always, the Unknown is perturbed by an amount Δ. Loads move in the lofted configuration and Equation 1.1 of Chapter 1 is applied. Yet rather than use actual loads, we use a unit load of 1 force/area, then the Unknown force of Equation 1.1 immediately identifies the area associated with that perturbation. The actual equation will be shown shortly.

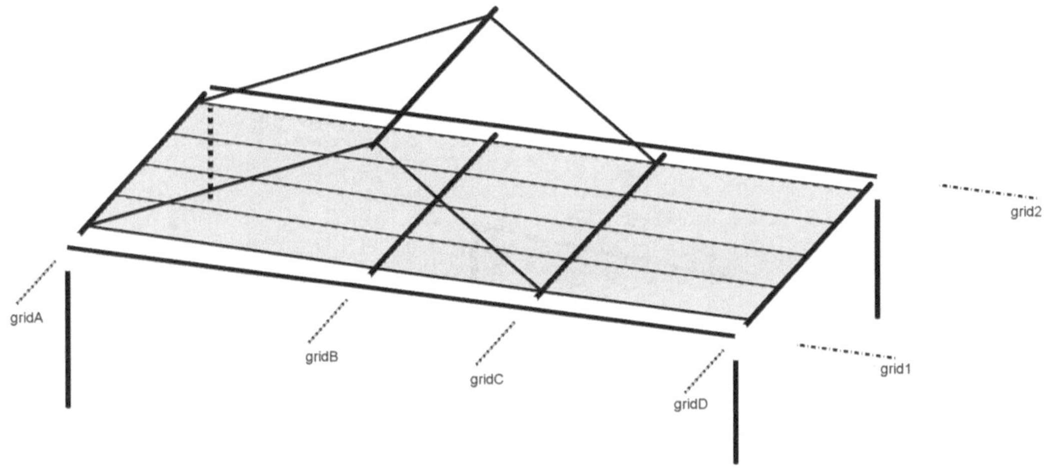

6.3
Perturbation of the entire beam on B

The technique in all of these horizontal grid problems is always the same, perturb the item being investigated, apply a unit force per area, find the lofts of the centroid of the *original* shape, and identify the tributary area that flows into the element being investigated. For example, if we seek the tributary area of the beam on grid line B of Figures 6.2 and 6.3, we first establish the centroid of the floor areas being lofted, ensuring that if a kink is present, we break up the area into sub-areas on either side of the kink. Figure 6.4 shows the perturbation of 1, greatly exaggerated and lofts, here 0.5 and 0.5, for the tributary area associated with beam B.

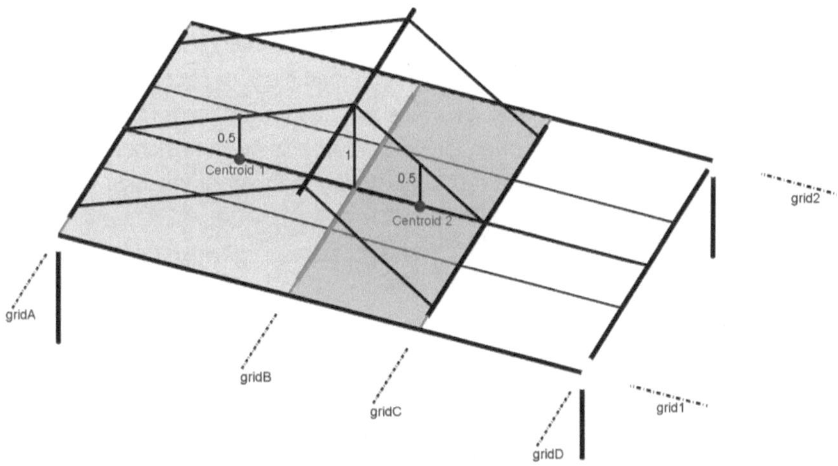

6.4
Establishing tributary area of the beam on B

Then the application of Equation 1.1 of Chapter 1, gives the force for any DL and LL, but really gives the tributary for the Unknown being studied, as:

$$Unknown\ Trib\ Area \cdot \Delta - 1\frac{force}{area}\cdot area1 \cdot Loft1 - 1\frac{force}{area}\cdot area2 \cdot Loft2 = 0 \quad \text{(Equation 6.1)}$$

Referring back to Figure 6.1, suppose we were asked to find the internal bending moment in the beam along grid line A, at a cut that is halfway between grid lines 2 and 3, i.e. at a cut 22.5 units of length north of the column at A1. This will be done with DL and LL being accounted for separately, for those readers who would like to apply LL reduction which always flows to an element, and never to a particular cut in an element. But no such reduction will be used in these examples. The cut of interest is shown in Figure 6.5.

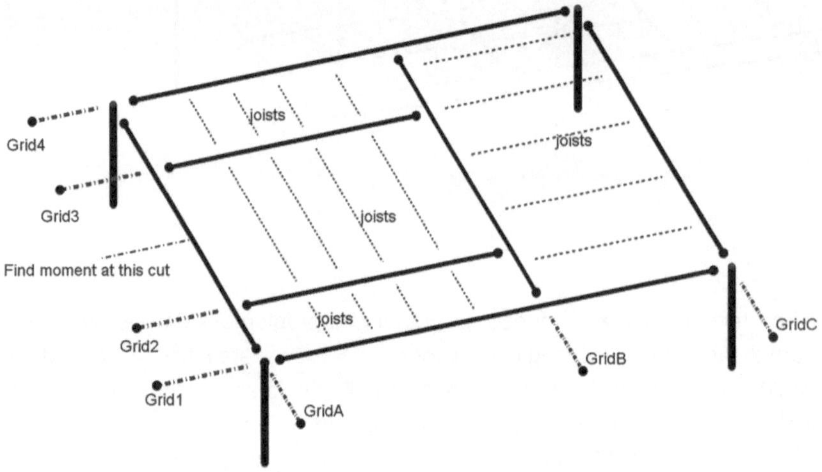

6.5
Ultimately seeking the magnitude of the moment at this cut

Figure 6.6 highlights the connection that is typical in Figures 6.1 and 6.2, for example, the beams on grid lines 2 and 3 connect into the beam along grid line A, but they do so without transferring moment. One way of doing this quickly in the field is with a "hangar" connection that is essentially a small bracket which vertically supports the lower hierarchical member. It is like a small "shelf" connected to a primary member, onto which rests the lower hierarchical, or secondary member. That "shelf" or hangar then pushes down as a concentrated point force, onto the primary member. Even if the connections are not precisely as sketched here, this is exactly what is meant by load flow. Load from the lower hierarchical member

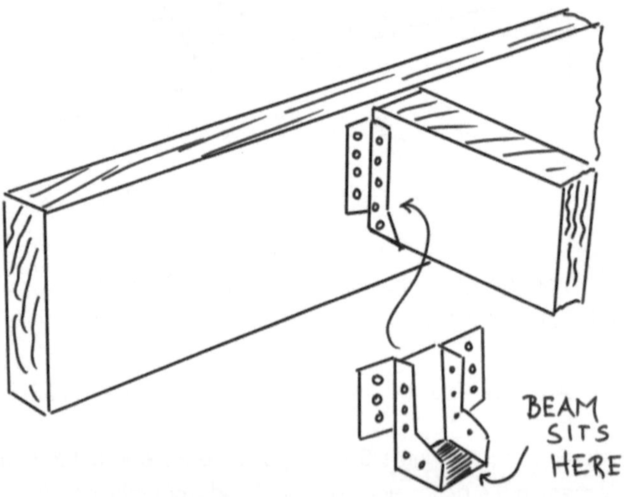

6.6
Simple hangar connection

flows or transfers to the higher hierarchical member. But that load is always a concentrated downward force in timber construction. A bending moment is never transferred in timber construction, and this is also the case in many steel horizontal grids as well.

Study of the small gaps between elements in all the drawings and re-emphasizing the fact that only a downward force is transferred in the typical connection leads us to conclude that the beam (or girder) on grid line A in Figure 6.5 experiences two point loads, one coming from the beam on grid line 2 and one coming from the beam on grid line 3. Notice that the beams on grid lines 1 and 4 do not touch the beam on grid line A. Those two beams connect directly to the columns, and load flows from them into the columns, not to the beam on grid line A. The reader should also notice that the joists in Figure 6.1 do not touch the beam on grid line A, thus load from the joists cannot flow into the beam on grid line A. Only the two concentrated forces flow into the beam on grid line A.

That critical idea must be understood before we establish the quantity of force that flows from the connection of the beam on 2 to the beam on A and similarly, the connection from the beam on 3 to the beam on A. To actually quantify this concentrated downward force, which flows through a connection such as was shown in Figure 6.6, apply the Müller-Breslau Method to the problem. In Figure 6.7, the force in the connection between the element on 2 and the element on A is the unknown being sought. The perturbation is vertical since we are only dealing with vertical load flow through that connection. The magnitude of this perturbation Δ is purposely large so the reader will see the lofted surface clearly. Yet even with a very large Δ, the answer from Equation 1.1 in Chapter 1 will be exactly correct, without the use of small perturbations. This is so because we only have gravity applied loads in this problem, thus the Classical Müller-Breslau Method will be used here, not the Modern Müller-Breslau Method. As stated elsewhere, the Classical Müller-Breslau Method stretches elements that are lofted.

Here is Equation 1.1 from Chapter 1 repeated for the unknown force in the hangar connection.

$$\textit{Vertical Connecting Force} \cdot \Delta + \sum \textit{Force}_i \cdot \textit{loft}_i = 0 \qquad \text{(Equation 6.2)}$$

Recall that when a kink in the lofted surface is formed, the lofts must be established separately on either side of the kink. This is the case in Figure 6.7 which shows a large Δ at the cut hangar connection. Since we divide by Δ in Equation 1.1 to find the unknown force, the magnitude of Δ is irrelevant for gravity only loading as we have here. Alternatively, one can think of Δ as being 1, and that the Δ in Figure 6.7 is largely exaggerated to emphasize the loft.

To demonstrate that this method can indeed be applied to a simple table napkin sketch, consider the similar triangles sketched out in Figure 6.8.

Applying Equation 1.1 to this problem immediately finds the unknown vertical force, known to experienced readers as "shear". But an even more interesting insight is that we can apply a unit load of 1 force/area, instead of the given DL and LL in the problem, and in so doing we can find the area that "flows into" that connection between the two elements.

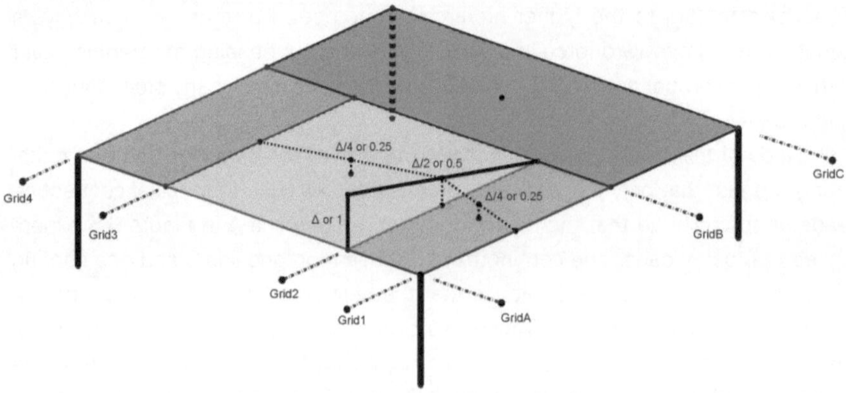

6.7
Finding area associated with force at A2 connection

$$UnknownArea \cdot 1 \, {}^{force}\!/_{area} \cdot \Delta - 1 \, {}^{force}\!/_{area} \cdot AB12 \cdot \frac{\Delta}{4} - 1 \, {}^{force}\!/_{area} \cdot AB23 \cdot \frac{\Delta}{4} = 0$$

(Equation 6.3)

$$UnknownArea = 262.5 \; length^2$$

(Equation 6.4)

This means that an area of 262.5 length^2 flows into the connection between the beam on 2 and the girder on A. This area is known as a "tributary area", much as in hydrology a "tributary area" drains from a specific watershed. Here, the tributary area flowing through that specific connection is TribA2 = 262.5 length^2.

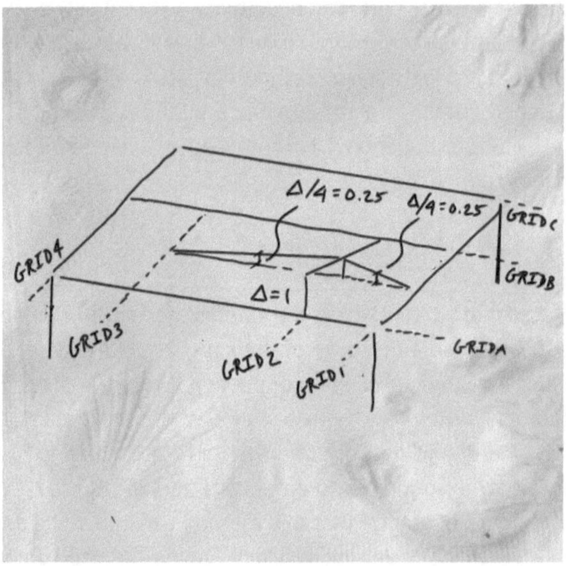

6.8
Using a sketch to find area flowing into A2 connection

Figure 6.9 investigates the second connection, by the same technique. We seek the new tributary area that flows into the connection at A3.

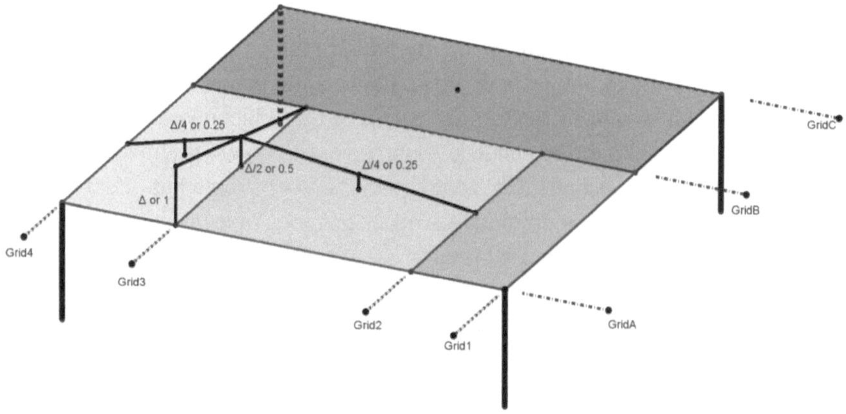

6.9
Finding area associated with force at A3 connection

To show that all these sketches can be quickly done without a computer, Figure 6.10 finds the lofted configuration with a table napkin sketch.

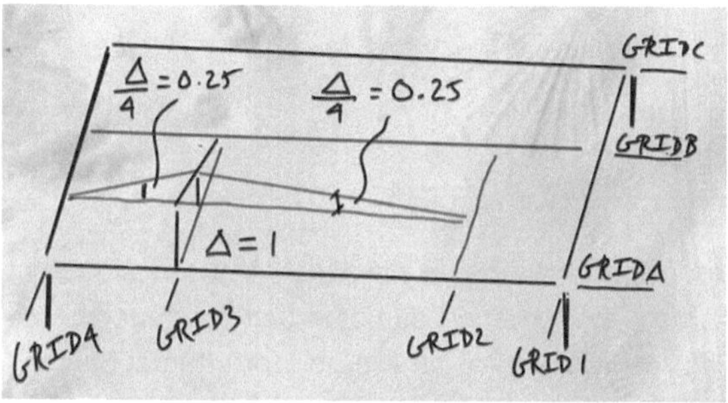

6.10
Finding area associated with force at A3 connection: table napkin sketch

$$UnknownArea \cdot 1 {}^{force}\!/_{area} \cdot \Delta - 1 {}^{force}\!/_{area} \cdot AB23 \cdot \frac{\Delta}{4} - 1 {}^{force}\!/_{area} \cdot AB34 \cdot \frac{\Delta}{4} = 0$$ (Equation 6.5)

$$UnknownArea = 277.5\ length^2$$ (Equation 6.6)

This means that a tributary area TribA3 = 277.5 length^2 flows into the connection between the beam on 3 and the girder on A.

Having the tributary areas associated with each of the two connections is a wonderful tool since we can now use these areas for both Dead Load (DL) and for Live Load (LL) separately. This would allow for any LL Reduction, which will not be used in this example.

The force flowing through the connection associated with the Dead Load is simply the Dead Load (force/area) multiplied by the tributary area. The force flowing through the connection associated with the Live Load is the Live Load (force/area) multiplied by the tributary area. The total force at each connection to the girder on A is the superposition of DL and LL.

What becomes absolutely clear in Figures 6.7–6.10 is that the entire area between grid lines B and C has absolutely no influence on the forces flowing into beam A. The Müller-Breslau Method makes this visually apparent because that entire area east of grid line B has no loft at all due to the perturbations.

As stated previously, the DL and the LL flowing into beam A are shown separately, in case the advanced readers want to use LL reductions. In Figure 6.11, the arrows representing the four individual forces are all drawn to the same scale. Drawing multiple applied loads to a single scale is a very simple but effective tool which helps the designer understand relative magnitudes of loads and what might be more influential, what might be less so.

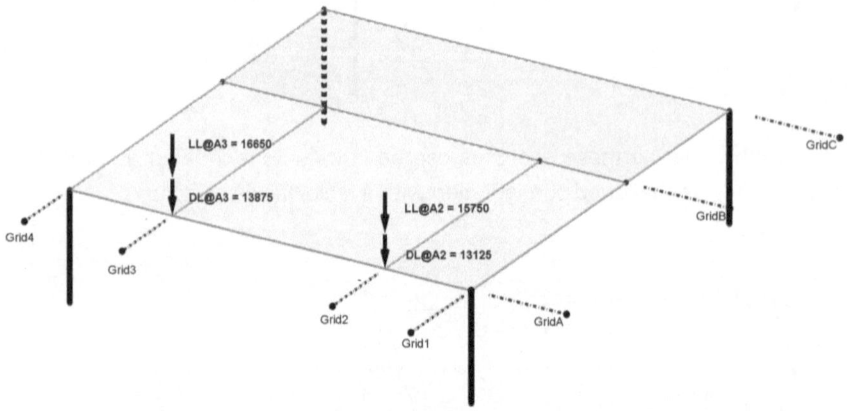

6.11
Forces arising from DL and LL drawn to the same scale

To calculate the internal moment in the beam on grid A, the techniques of Chapter 3 are used to establish the perturbation due to a unit crack at the cut of interest. Specifically, the Classical Müller-Breslau Method for internal bending moment, shown in Figure 2.28 of Chapter 2 is used here. The height of the cracked cut is determined by the span left of the crack, known as the "a segment" and the span right of the crack, known as the "b segment". The technique of Chapter 2 is restated here, with the height of the crack of Equation 6.7 here being 11.7 units of length, as shown in Figure 6.12, with the application to our problem at hand described in Figure 6.13.

$$Ht@Crack = \frac{a \cdot b}{a+b} = \frac{24.5 \cdot 22.5}{24.5+22.5} = 11.7 \qquad \text{(Equation 6.7)}$$

Notice in Figure 6.12, that although the Δ perturbation of Equation 1.1 is the relative angle on either side of the crack, an easy way to establish such a unit angle, and consequently such a unit Δ, is to find the height of the crack in the perturbed configuration, i.e. to use the previously defined Ht@Crack from Equation 6.7. This is the preferred technique when using the Classical Müller-Breslau Method.

Figure 6.14 shows the same geometric solution on a hand sketch.

Here, the internal bending moment is the unknown of Equation 1.1 of Chapter 1, and the forces and the lofts are clearly shown in Figures 6.13 and 6.14. Thus:

$$UnknownMoment \cdot 1 - Force@A2 \cdot Loft@A2 - Force@A3 \cdot Loft@A3 = 0 \qquad \text{(Equation 6.8)}$$

$$Unknown\ Moment = 326546\ force \cdot length \qquad \text{(Equation 6.9)}$$

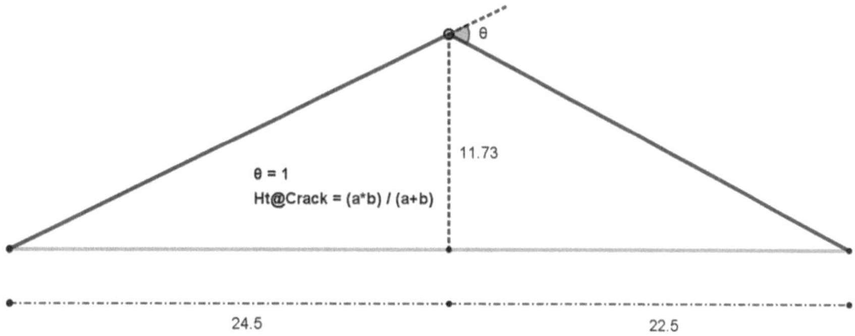

6.12
Classical Müller-Breslau Method perturbation to find the internal bending moment

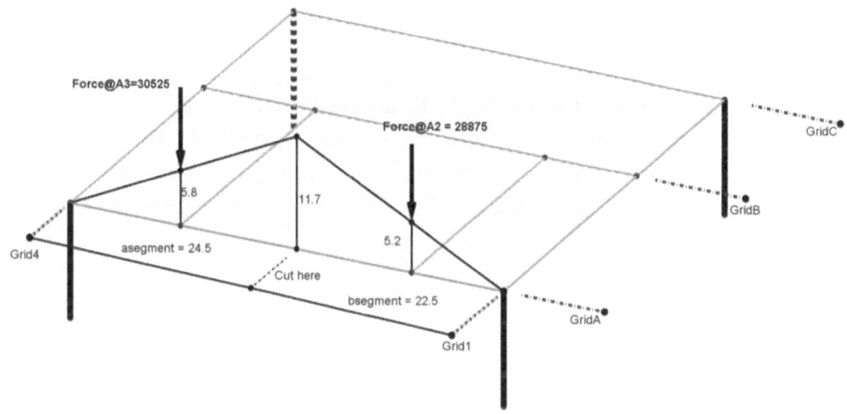

6.13
Finding internal bending moment at the cut of interest

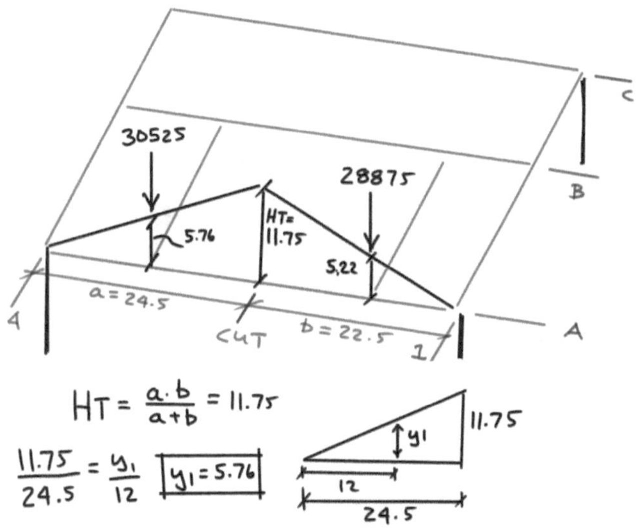

6.14
Finding internal bending moment at the cut of interest by hand sketch

The configurations of Figure 6.15, 6.20, 6.26 and 6.30 were originally studied by T. Bart Quimby on his website bgstructuralengineering.com. They have been solved in a completely novel way herein.

Consider now another example where the joists change directions. Notice that there are five columns in the plan view, shown in Figure 6.15, and four primary beams. We will call the x direction moving east, and the y direction moving north. Joists run

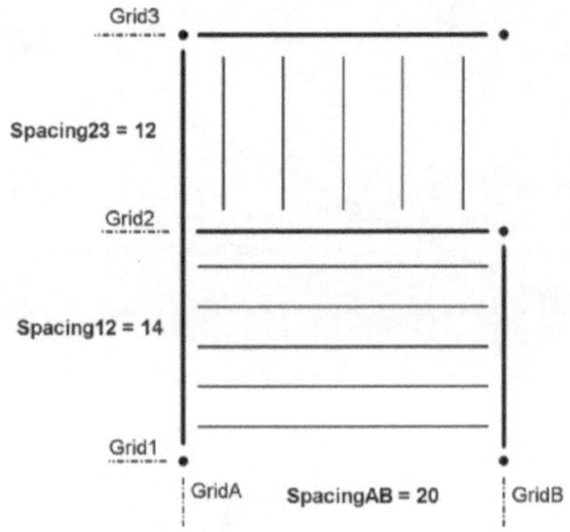

6.15
Example of the horizontal grid of beams shown in 2D plan view

north/south between grid lines 2 and 3. Joists run east/west between grid lines 1 and 2. Here, we seek the tributary area associated with the column at A3. This area could be used for Live Load Reduction. However, in this example, neither the Dead Load nor the Live Load is given; we simply seek the tributary area of column A3.

For more clarity, Figure 6.16 shows the same horizontal grid of beams, supported by pin-ended columns in a 3D view.

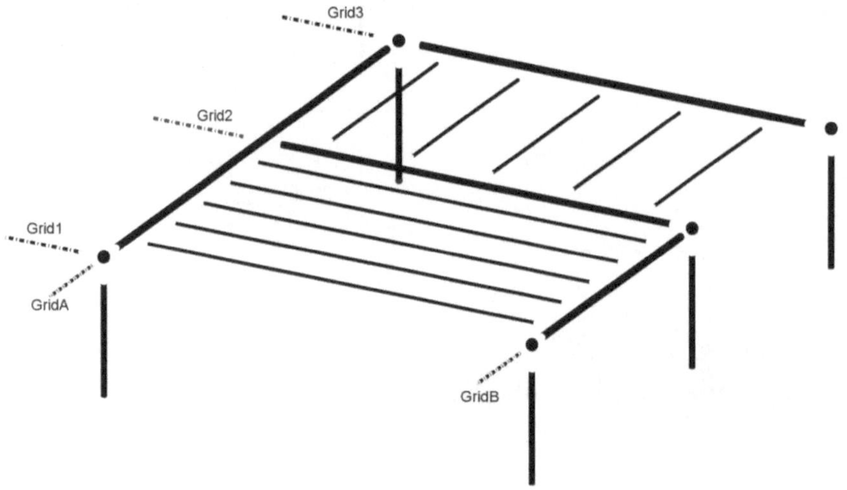

6.16
Example of the horizontal grid of beams shown in 3D view

Recall from previous examples, that for gravity loads only, we can loft Δ to very large values using the Classical Müller-Breslau Method, and the answer will be exact. Here, the unknown is the area associated with the force in column A3. But to find that area, we imagine that a unit floor load of 1 force/area is applied uniformly. Then, since there are no kinks in the lofted surface, we only need to find the centroid of the original floor quadrilateral. Then, never ever go directly from the column of interest to that centroid! Always go down mutually perpendicular slopes, typically east/west or north/south, but always along straight edges of the lofted surface. Here, similar triangles establish two lofted values, this is shown in Figure 6.17. There is nothing special about the Δ value of 6, it was chosen to be large and visible. A Δ of 1 could have also been used. Notice the centroid of the original floor quadrilateral is located halfway between the grid lines A and B which are 20 units of length apart, and halfway between the grid lines 1 and 3 which are 26 units of length apart. This centroid lies just slightly south of grid line 2.

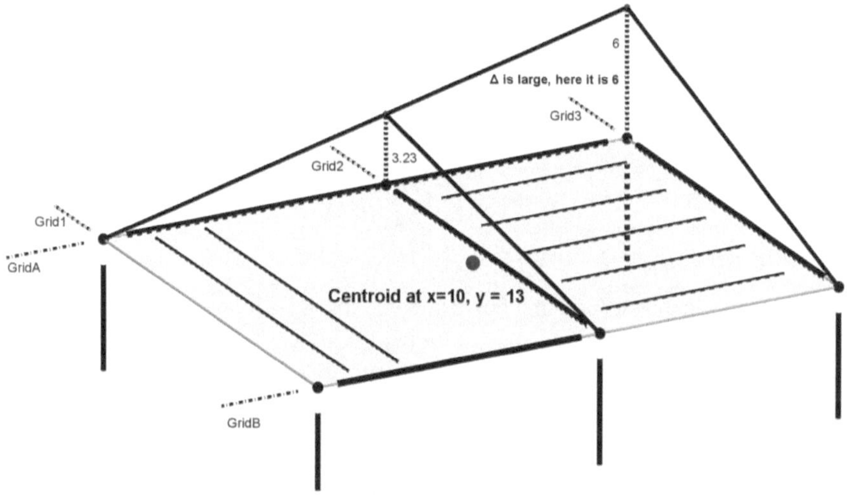

6.17
Example of the horizontal grid of beams shown in 3D

Since the entire centroid of the floor is at $x = 10$ and $y = 13$, the loft at that point must be found. Using only east/west or north/south movements, as well as similar triangles, the loft is immediately found as 1.5. This is shown in Figure 6.18. Notice that the loft 1.5 is exactly 0.25 of the original loft of 6 at column A3.

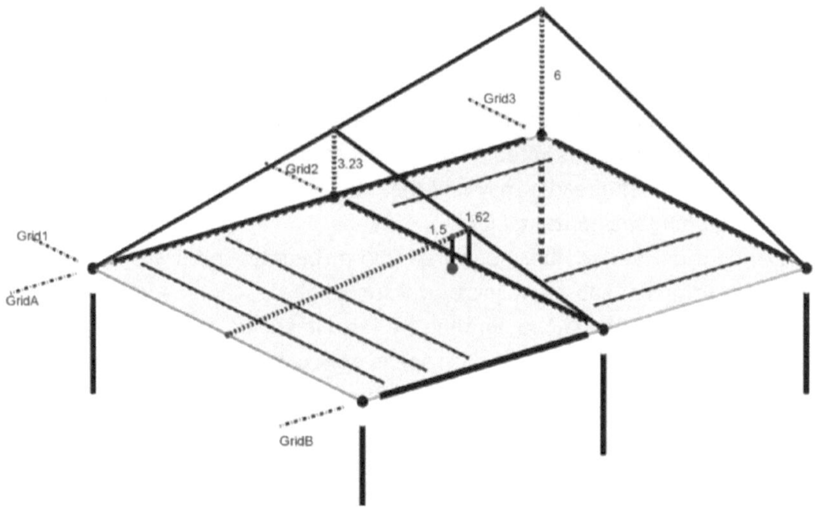

6.18
Finding the tributary area associated with column A3

Figure 6.19 shows the exact same geometric construction on a hand sketch, but for a unit loft of 1 at column A3. The value of 0.25 corroborates the previous digital drawing.

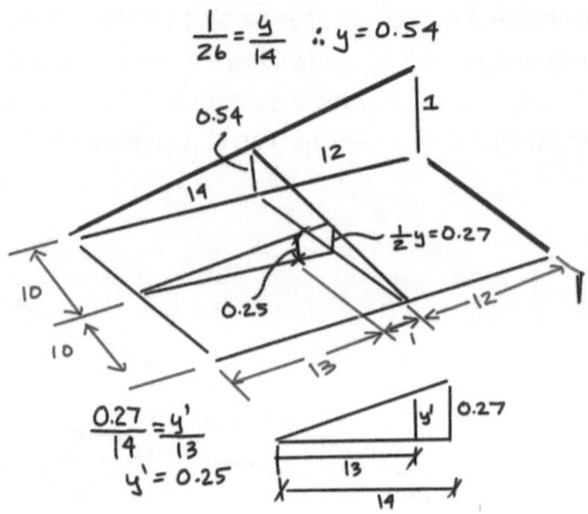

6.19
Finding the tributary area associated with column A3 via hand sketch

Then the Unknown Area that flows into column A3 is found via Equation 1.1 of Chapter 1. All values are known except for the Unknown Area.

$$Unknown\ Area \cdot 1\ {}^{force}\!/\!_{area} \cdot \Delta - 1\ {}^{force}\!/\!_{area} \cdot area\ of\ AB13 \cdot LoftAB13 = 0$$ (Equation 6.10)

$$Unknown\ Area = 130\ length^2$$ (Equation 6.11)

Clearly, the unit force per area cancels in the above equation, making it quick to see the efficiency of the equation simply in terms of Unknown Area.

$$Unknown\ Area \cdot \Delta - area\ of\ AB13 \cdot LoftAB13 = 0$$ (Equation 6.12)

$$Unknown\ Area = 130\ length^2$$ (Equation 6.13)

Notice that the tributary area of Column A3 in Figure 6.15 is 130 length^2 and the total area is 520 length^2. Thus the ratio 130/520 is equal to ¼. Then notice that from Figure 6.19, the entire area is lofted 0.25, which instantly tells us that ¼ of the area flows to column A3. The reader may find that surprising, given that there are five columns supporting the horizontal grid!

Next, consider yet another horizontal grid of beams. There are 12 primary beams, nine columns, and many joists which change direction as shown in Figure 6.20. Neither the Dead Load nor the Live Load is shown, rather, we seek the tributary area of the primary beam which spans from column A1 on the north-west corner to column B2 which is somewhat centrally located. As always, the gaps in the plan view symbolize a simple, non-moment-transferring connection to the subsequent element.

The horizontal grid is shown in the 3D view in Figure 6.21.

To determine how much area "flows to" the beam A1-B2, i.e. to find this beam's tributary area, perturb the beam from its original state. In Figure 6.22, the

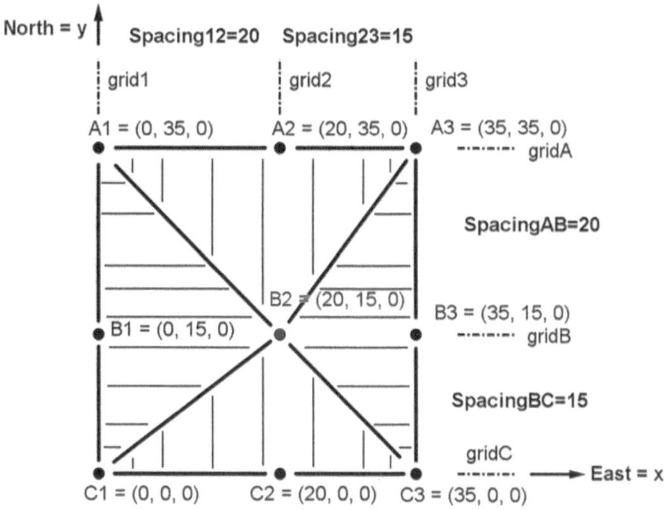

6.20
Finding the tributary area associated with beam A1-B2

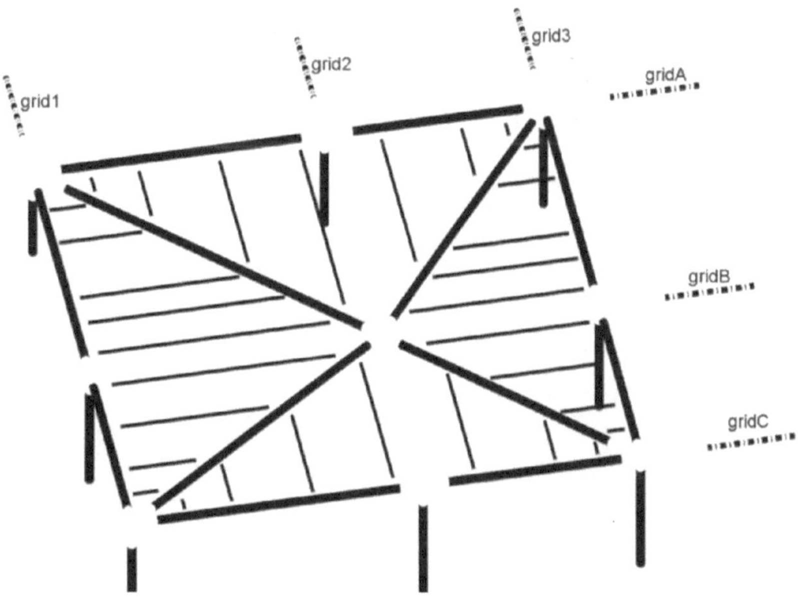

6.21
Finding the tributary area associated with beam A1-B2, 3D view

perturbation is large so as to be visible, here $\Delta = 8$ units of length. Notice that the joists then drop down from this perturbed state to zero along grid A and zero along grid 1, and that the sweep downwards is parallel to grid 1 on one side and parallel to grid A on the other side. Figure 6.22 shows the line sloping from Δ down to zero, parallel to grid A and B, which intersects with a vertical line through that triangle's centroid. The loft is the height of a segment vertically rising from the original centroid, to where the centroid would be elevated in the lofted configuration.

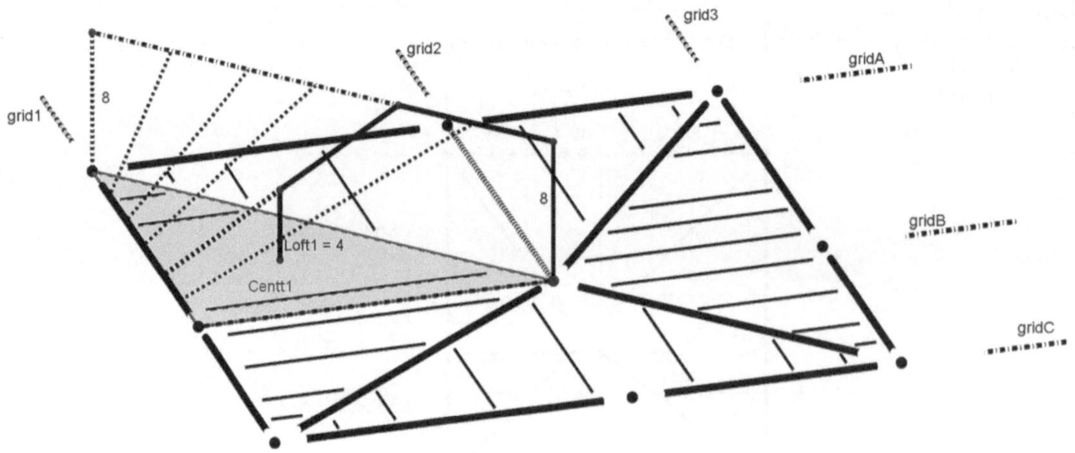

6.22
Perturbed beam A1-B2 creates Lofts: here Loft1 is shown

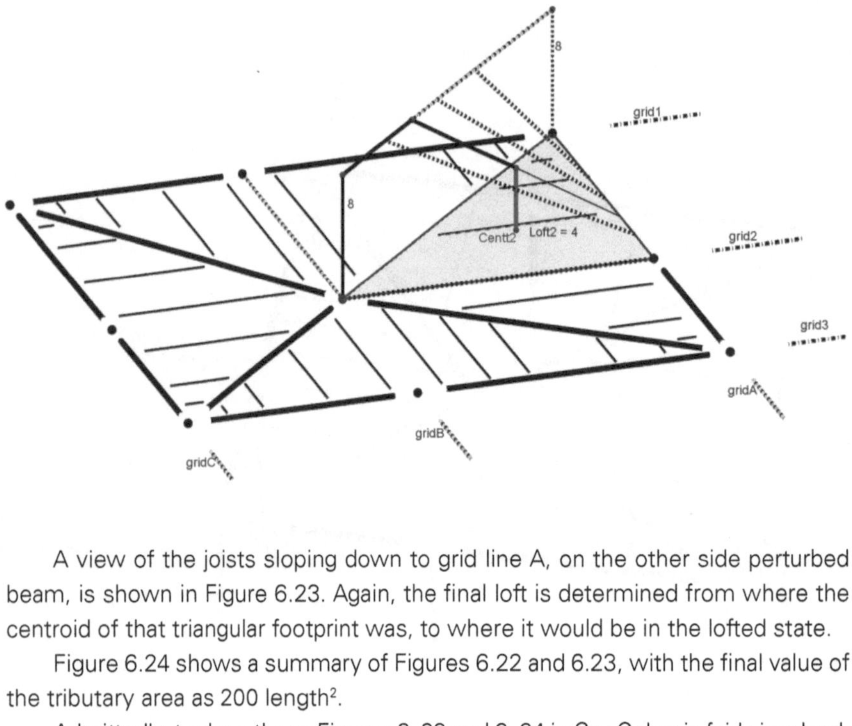

6.23
Perturbed beam A1-B2 creates Lofts: here Loft2 is shown

A view of the joists sloping down to grid line A, on the other side perturbed beam, is shown in Figure 6.23. Again, the final loft is determined from where the centroid of that triangular footprint was, to where it would be in the lofted state.

Figure 6.24 shows a summary of Figures 6.22 and 6.23, with the final value of the tributary area as 200 length^2.

Admittedly, to draw these Figures 6–23 and 6–24 in GeoGebra is fairly involved. This is a clear case where the hand sketch is far quicker to create. Figure 6.25 shows the sketch of the perturbed beam and the Loft on either side of the beam.

The Müller-Breslau Method is applied exactly the same way for each and every determinate, horizontal grid problem. Complexities arising from changing joist directions, from openings in the floor, even from non-uniform loading all fall away because the only thing that is needed is a few straight lines, always parallel to the edges of the lofted shape, never ever diagonal from one corner to the vertical line intersecting the centroid of the sub-area being lofted.

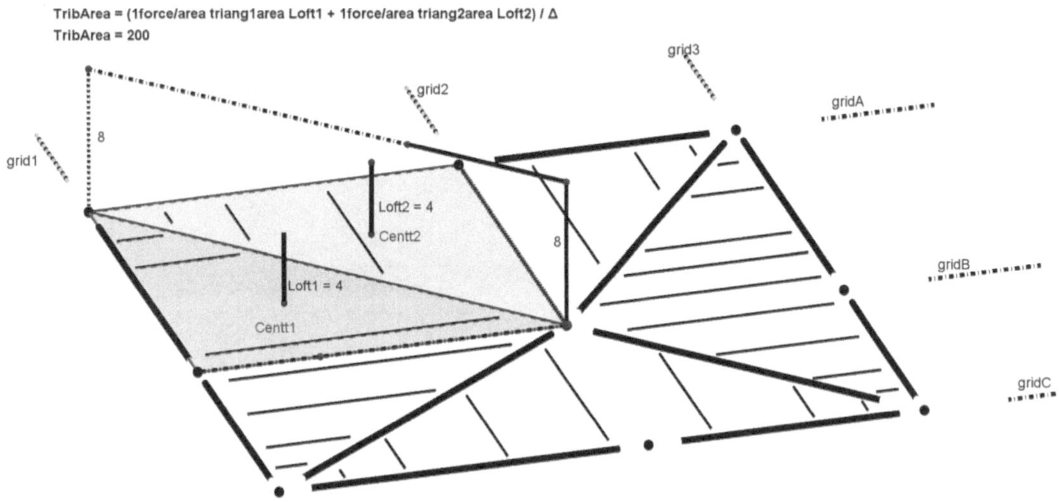

6.24
Perturbed beam A1-B2 creates Lofts: here both Lofts are shown

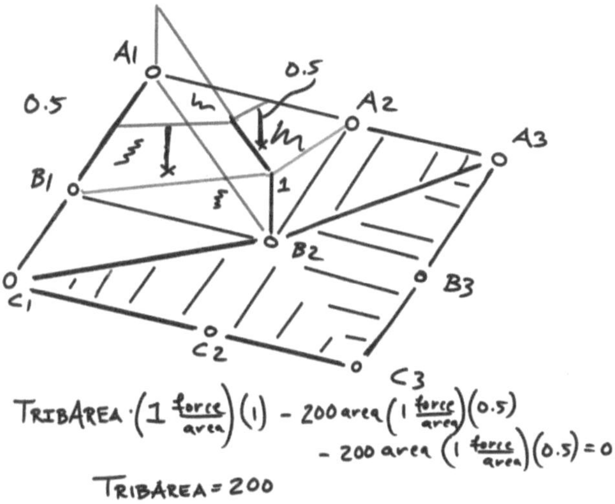

6.25
Perturbed beam A1-B2 creates Lofts: hand sketch

Consider a radial configuration of a determinate grid of beams, as shown in Figure 6.26. Suppose that we sought the tributary area that flows to the beam on grid line C.

Figure 6.27 shows the same setup as Figure 6.26 but in three dimensions. In Figure 6.26, notice that the builder chose to lay out the joists parallel to the leftmost primary beam in each section of the floor. Furthermore, notice that not all the joists between grid line B and grid line C flow into the beam on C being investigated.

Only the joists that connect to the beam on C actually get lofted during beam C's perturbation. As stated previously, not all the joists between B and C get lofted, and none of the joists between C and D get lofted at all. What is the area being lofted? A line parallel to grid line B, passing through the inner radius of the floor system immediately establishes the footprint that will be lofted. The technique is to find the centroid of that shape in the original, horizontal position, move along known edges of the perturbed beam and slide down parallel to any joists till the

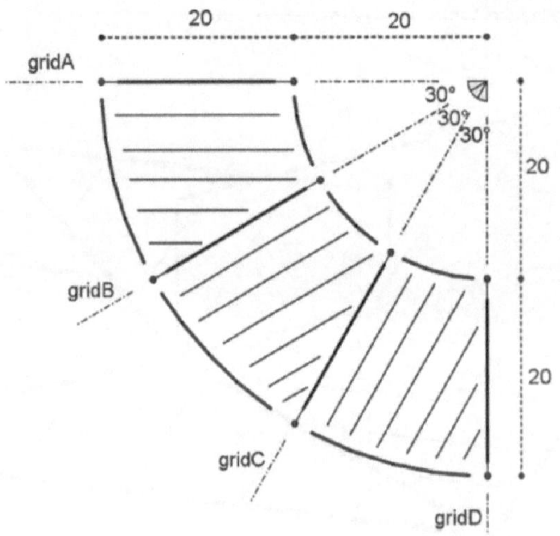

6.26
Plan view, radial configuration, find the tributary area of beam C

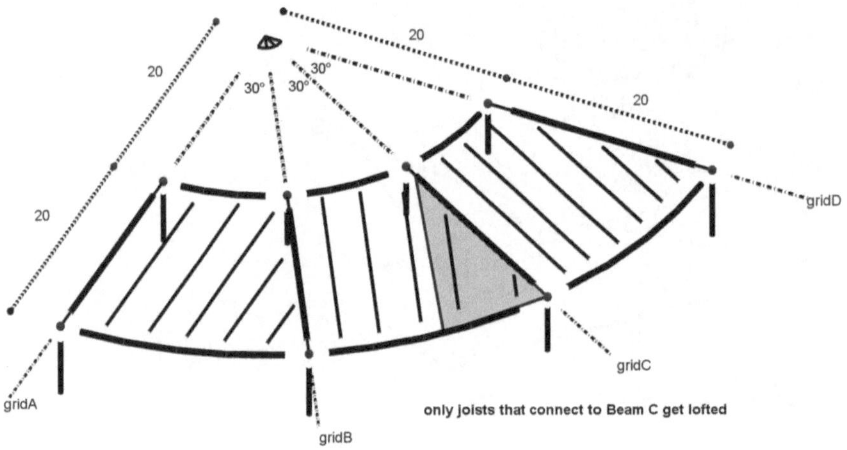

6.27
3D view, radial configuration, find the tributary area of beam C

loft is zero. Then find the intersection of the straight sloped line and a vertical line passing through the centroid. This is shown in Figure 6.28. The Δ in Figure 6.28 is set to a dramatically large value of 6 so that the lofts are clearly seen.

While Figure 6.28 is fairly complicated to draw on a computer, Figure 6.29 shows the same answer can be obtained with a quick hand sketch.

Equation 1.1 of Chapter 1 is used, as always, but here with a unit force/area, in order to establish the tributary area.

$$\textit{Unknown Area}\cdot 1\,{}^{force}\!/_{area}\cdot\Delta - 1\,{}^{force}\!/_{area}\cdot\textit{area lofted}\cdot\frac{\Delta}{2}=0 \qquad \text{(Equation 6.14)}$$

$$\textit{Unknown Area}\cdot\Delta-\left(110\ \textit{length}^2\right)\cdot\frac{\Delta}{2}=0 \qquad \text{(Equation 6.15)}$$

$$\textit{Unknown Area}=55\,\textit{length} \qquad \text{(Equation 6.16)}$$

Thus, the tributary area flowing to beam C is 55 length^2.

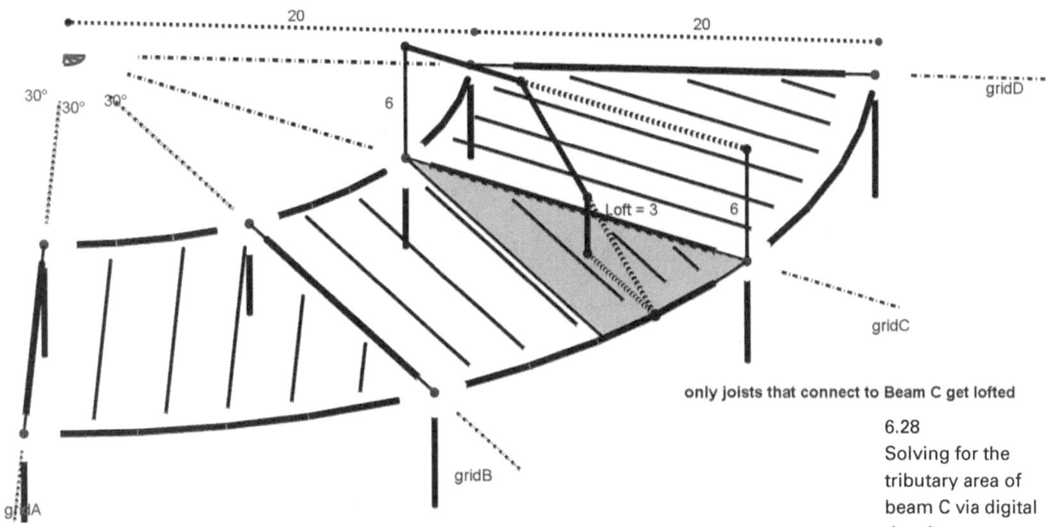

6.28
Solving for the tributary area of beam C via digital drawing

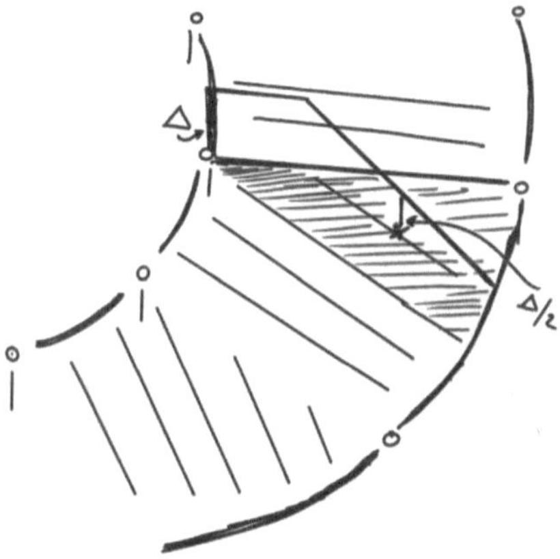

6.29
Solving for the tributary area of beam C via hand sketch

Figure 6.30 shows a different type of determinate configuration, one that has primary beams radiating from the column at B1. As in all other examples, the joists here are simply supported to the beams. We seek the tributary area which flows into column B1.

6.30
Plan view of the horizontal grid, seeking a tributary area of column B1

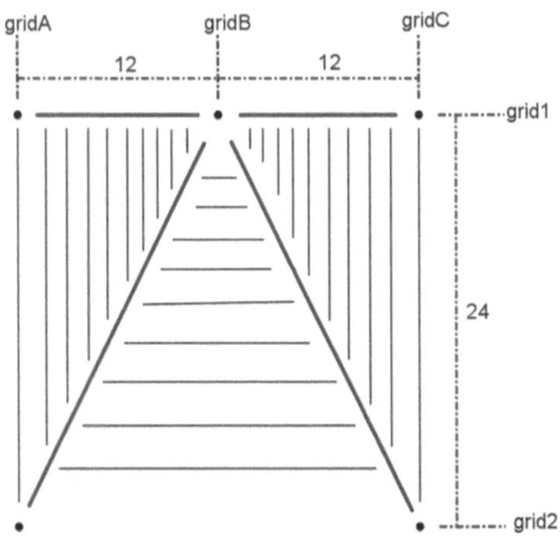

The same grid is shown in 3D in Figure 6.31.

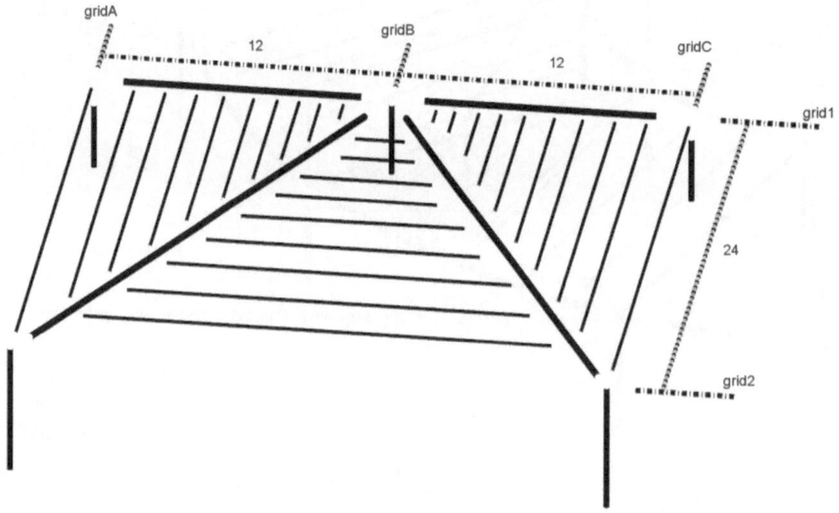

6.31
3D view of the horizontal grid, seeking a tributary area of column B1

As always, perturb the element of interest an amount Δ. Figure 6.32 shows the column at B1 perturbed a large amount.

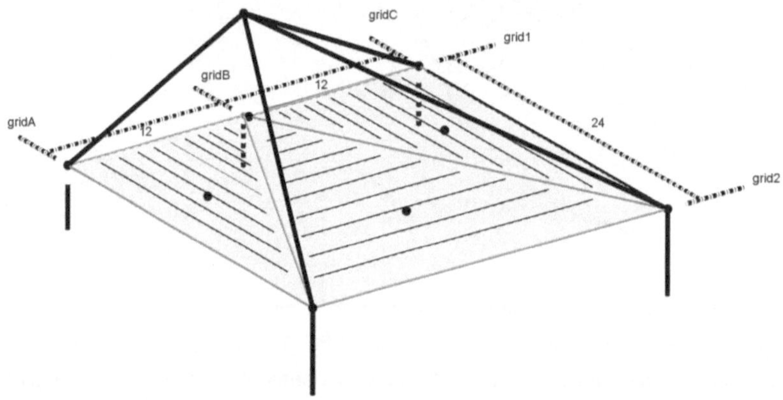

6.32
Perturbation of column B1

Although the grid seems a bit complicated, in reality, the fact that the centroids are those of triangular areas makes the problem quick to solve. The centroid lies at ⅓ the distance from the base of the triangle, thus, the path down the straight lofted members stops at ⅓ the distance from the end. All the sloped members have the same slope, so the problem is unlocked rapidly. This leads to the answer of ⅓ of the total area flowing into the column at B1. This is shown in Figure 6.33.

Figure 6.34 shows the hand sketch solution, which gives the same answer of ⅓ the total area as a tributary area to column B1.

As a final example that is again a bit different, Figure 6.35 shows a determinate horizontal grid of beams, but here a cantilever exists, east of column B1, noting the north direction arrow in Figure 6.35. The beam on grid line 1 supports the beam on grid line C with a simple connection. Notice that the beam on grid line 1 is primary,

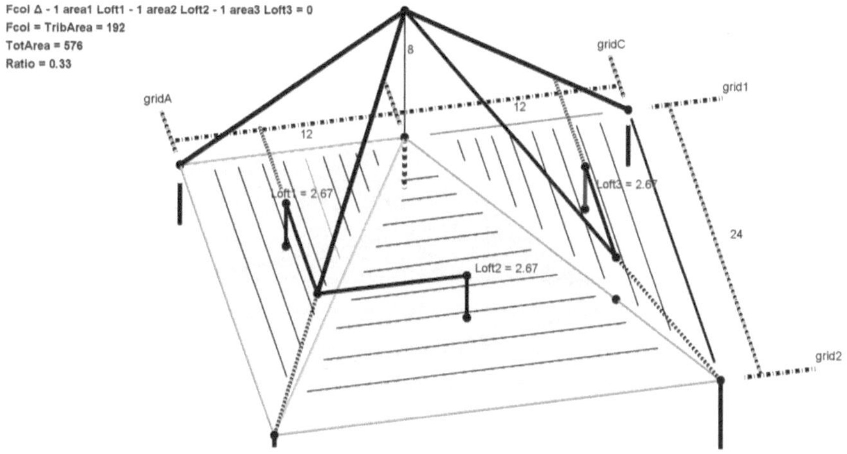

6.33
Digital drawing solution of the tributary area of column B1

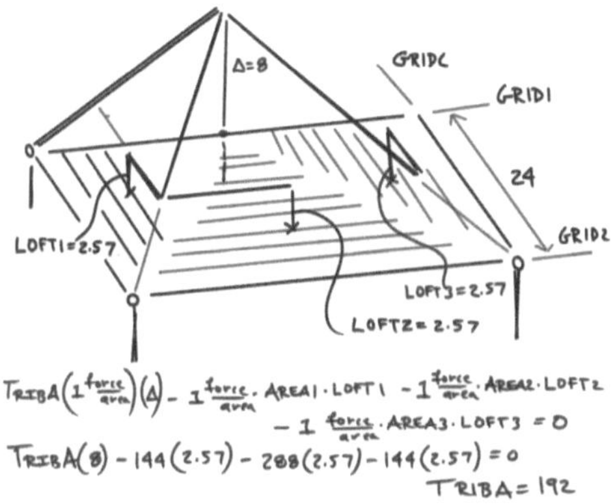

6.34
Hand-sketched solution of the tributary area of column B1

compared to the beam on grid line C. That beam on grid line C would be unstable without the connection to the primary beam since the connection of beam C to column C3 is not moment carrying. Finally, notice that the beam on the grid line1 passes over, or is stacked on top of, the column at B1. The reader should study Figure 6.35 and understand what is primary and what is secondary, before attempting to solve for any particular unknown. Notice that neither Dead Load nor Live Load is given in this problem, we will study only the areas that flow into certain elements. In particular, we seek the tributary area that flows into Column B1.

To find the tributary area which flows into Column B1, start by perturbing that column. We perturb the column at B1 a unit length. Figure 6.36 shows the lofted configuration of the floor for that unit perturbation. Notice that there are no kinks in the lofted shape, thus only one centroid for the overall floor is needed. Figure 6.36 immediately shows that the loft at that centroid is 0.33 for a unit Δ. Thus, for a unit distributed load of 1 force/area, the tributary area associated with the column on grid line B1 is found to be 64 length^2. This is because we found the force to be 64 units of force for the unit distributed load using Equation 1.1 of Chapter 1.

If we were interested in applying a live load reduction for the beam on grid line 1, we would need to establish the tributary area that flows into this beam. As such,

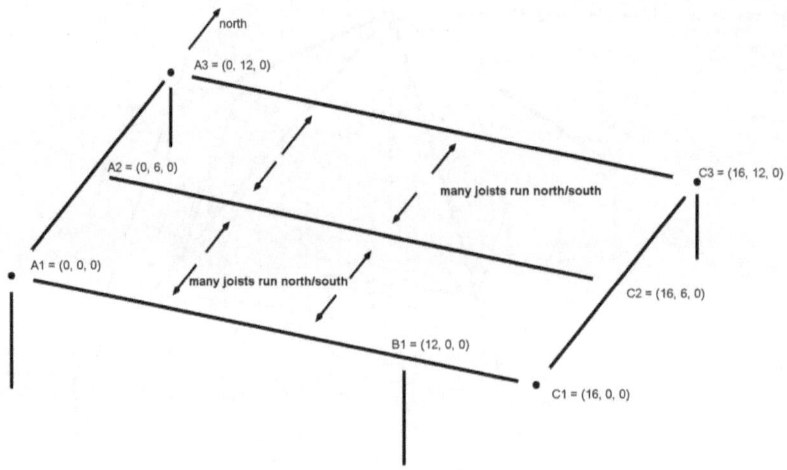

6.35
Horizontal grid of beams with one cantilever

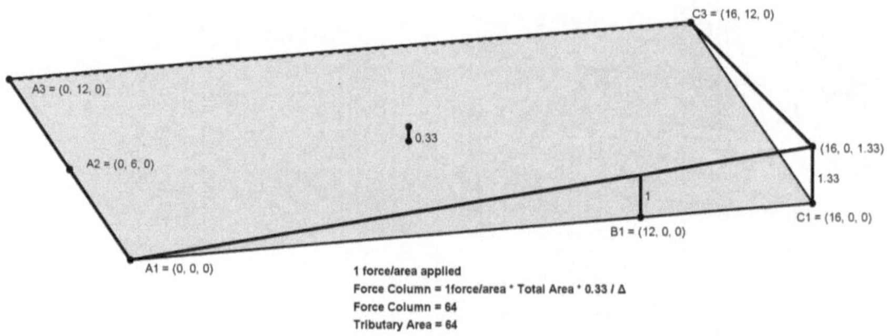

6.36
Establishing the tributary area of column B1

the entire beam is perturbed. Note that the joists running north/south would be lofted up along with the entire beam on grid line 1, but the beam on grid line A is not lofted, it connects to column A1 and to column A3, and not to the perturbed beam shown in Figure 6.37. The reader is encouraged to make a sketch of the loft induced by this perturbation, prior to looking at the solution which is shown in Figure 6.38.

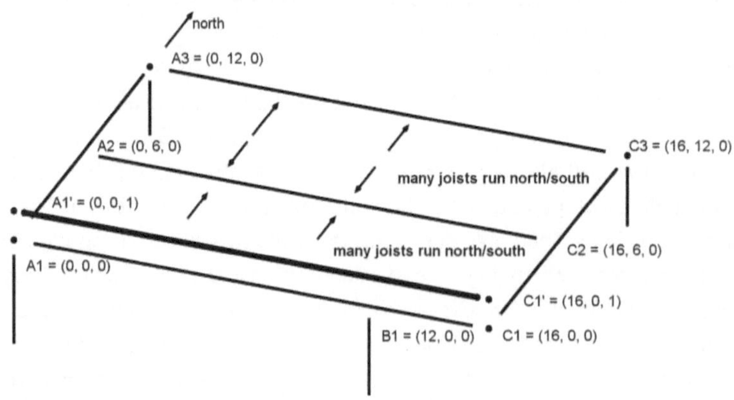

6.37
Perturbed beam along grid line 1

A kink is formed on the lofted surface, thus two centroids must be established, one on either side of the kink. Then, the application of Equation 1.1 of Chapter 1, using a unit force/area allows for the immediate solution of the area associated with the beam on grid line 1 as 72 length^2. The kink is shown in Figure 6.38.

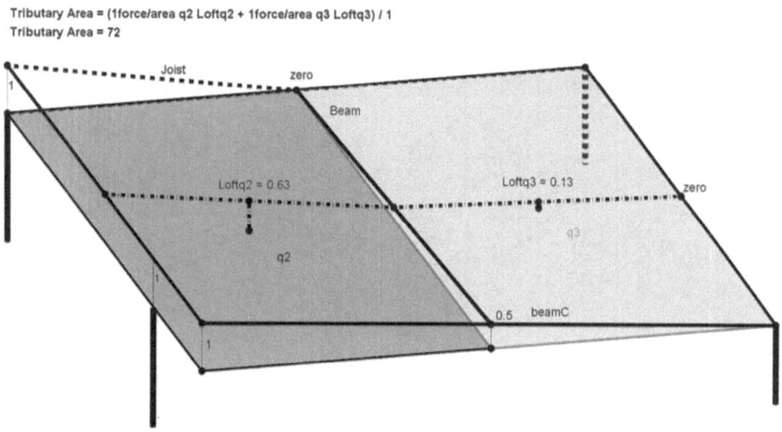

6.38
Tributary area of the beam along grid line 1

HORIZONTAL GRID ÉTUDES

The following studies or études should be sketched by hand.

Étude 6.1) The horizontal grid shown in plan view in Figure 6.39 will be used by several études. Note that the joists are not shown, but they are framed as shown (north/south). Note also the opening between grid lines 2 and 4 and B and C. Sketch it out in a 3D view, with some uniform column height at A1, A4, C1, and C4.

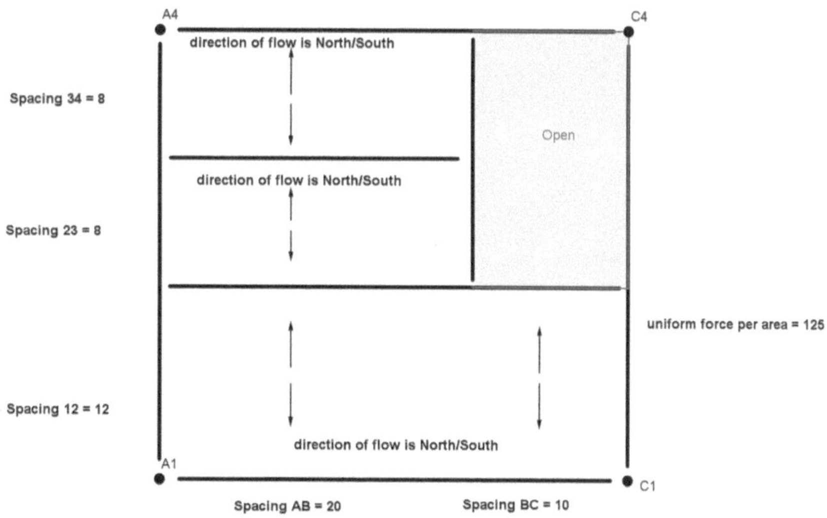

6.39
Étude 6.1 solution

Étude 6.2) For the grid of Figure 6.39, find the equilibrating axial force in column A1. But break all of the floor load into three discrete parts:

- part 1 from A1 to C2
- part 2 from A2 to B3
- part 3 from A3 to B4.

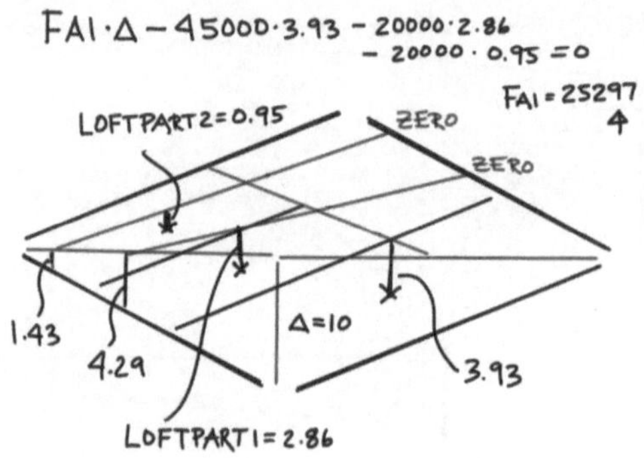

6.40
Étude 6.2 Solution

A hand sketched solution is shown in Figure 6.40.

Étude 6.3) Repeat Étude 6.2 but this time use only two subdivisions of the total load:

- part 1 from A1 to C2
- part 2 from A2 to B4.

A hand sketched solution is shown in Figure 6.41.

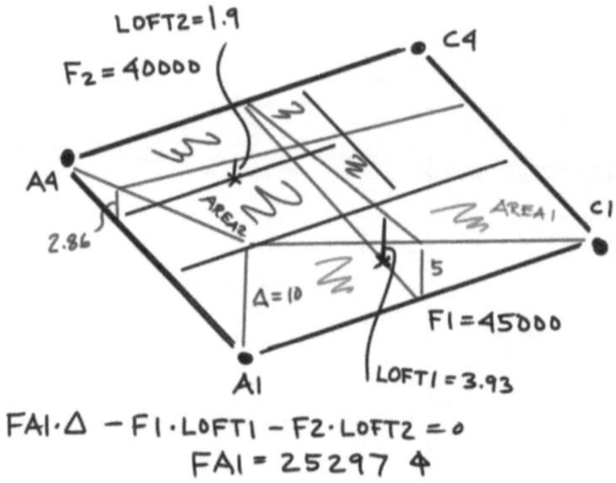

6.41
Étude 6.3 solution

Étude 6.4) Describe why we cannot simply use one point load to capture the entire floor load.

Answer: there is a kink in the shape along grid line 2. This is caused by the fact that the lofted line on grid line 3 has a different slope than the lofted line on grid line 2.

Étude 6.5) For the grid of Figure 6.39, establish the tributary area of the beam spanning between A2 and C2. A hand sketched solution is shown in Figure 6.42.

6.42
Étude 6.5 solution

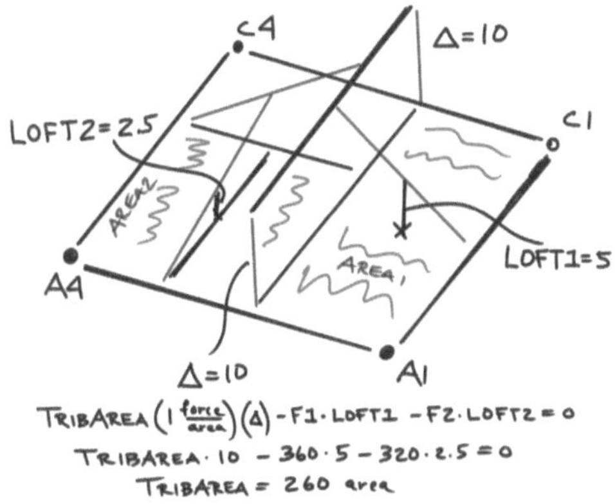

Étude 6.6) For the grid of Figure 6.39, establish the internal bending moment at the midspan of the beam spanning between A2 and C2. A digital sketch is shown in Figure 6.43. Use the technique of enforcing the Height of the Crack where the moment is sought to be strictly defined by the Classical Müller-Breslau Method and re-stated in Figure 6.12 as $HtCrk = \frac{a \cdot b}{a+b}$.

6.43
Étude 6.6 solution

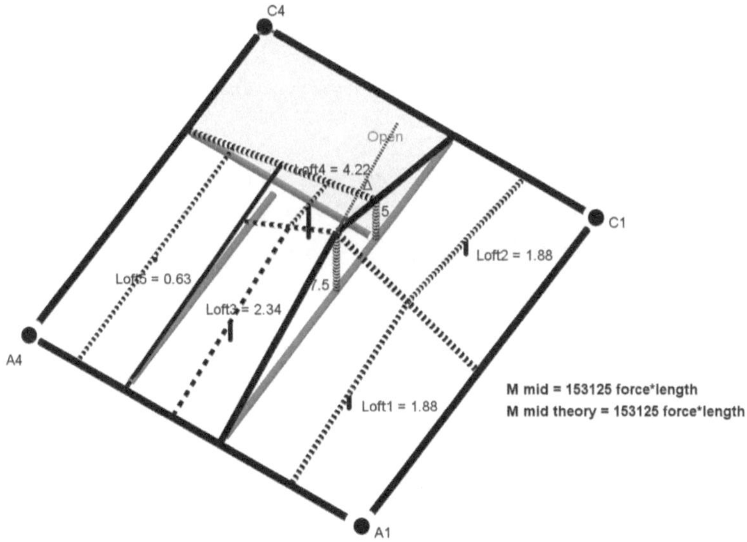

Étude 6.7) The horizontal grid shown in Figure 6.44 will be used by several études. Note that the joists are not shown, but they are framed as shown (north/south and east/west). Sketch it out in a 3D view, with some uniform column height at A1, A3, C1, B2, and C3.

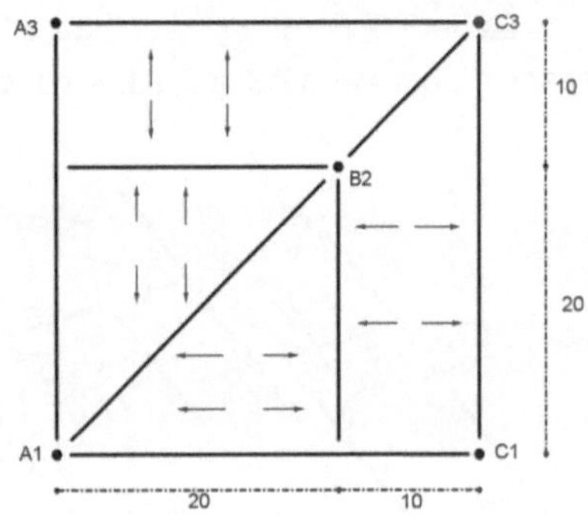

6.44
Étude 6.7 plan view

Étude 6.8) Given a uniform floor load of 100 force/area, find the axial equilibrating force in Column B2 in Figure 6.44. A hand sketched solution is shown in Figure 6.45.

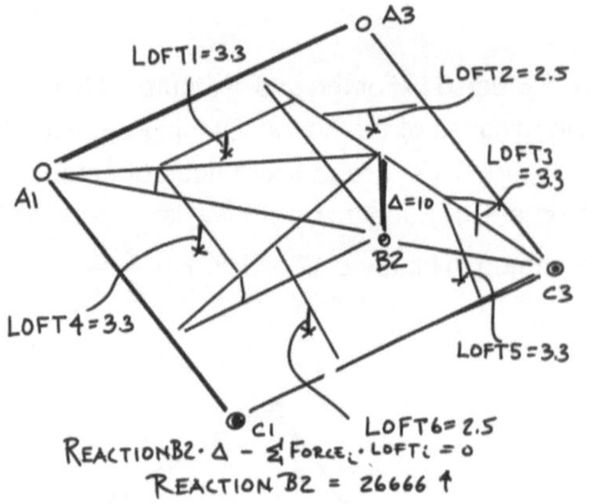

6.45
Étude 6.8 solution

Étude 6.9) Given a floor load of 100 force/area, find the internal bending moment in the mid-span of the beam between A1 and B2 in Figure 6.44. Do this for a $\Delta = 1$. A solution is shown in Figure 6.46.

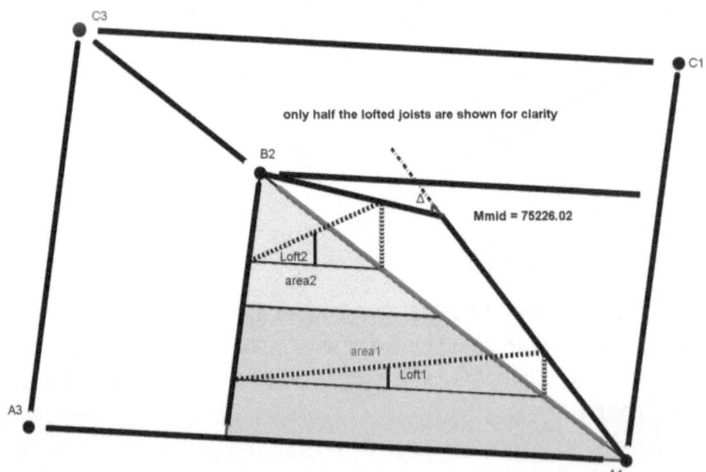

6.46
Étude 6.9 solution

7 Arches

So-called "single story skeleton structures" are really three-dimensional forms, but they are designed and analyzed in two dimensions when looking at them in "elevation view", i.e. as if one were on the floor of a large building, looking up to see magnificent arches holding up the roof. We will classify these large span forms as "arches" in this chapter, although other names could be used to describe them, such as gabled frames, or pitched frames. There is clearly much to learn from truss behavior, but notice the similarities and the major differences between any truss which was studied previously, and an arch such as the one shown in Figure 7.1.

As in truss structures, Figure 7.1 shows pin-ended elements, i.e. no moment can be transferred from the elements to the foundation, nor can moment be transferred from element to element at the crown of the structure. That connection is an internal hinge. Another similarity is that both arches and trusses can be readily studied in a 2D elevation view. However, what is different about Figure 7.1, compared to any truss, is that arches can be loaded along the member lengths themselves, thus inducing bending in individual members. Trusses are idealized to carry loads only at their connecting points, known as "nodes", and no bending occurs of any individual member, even as the entire truss deforms like a beam overall.

7.1
2D elevation view of the arch

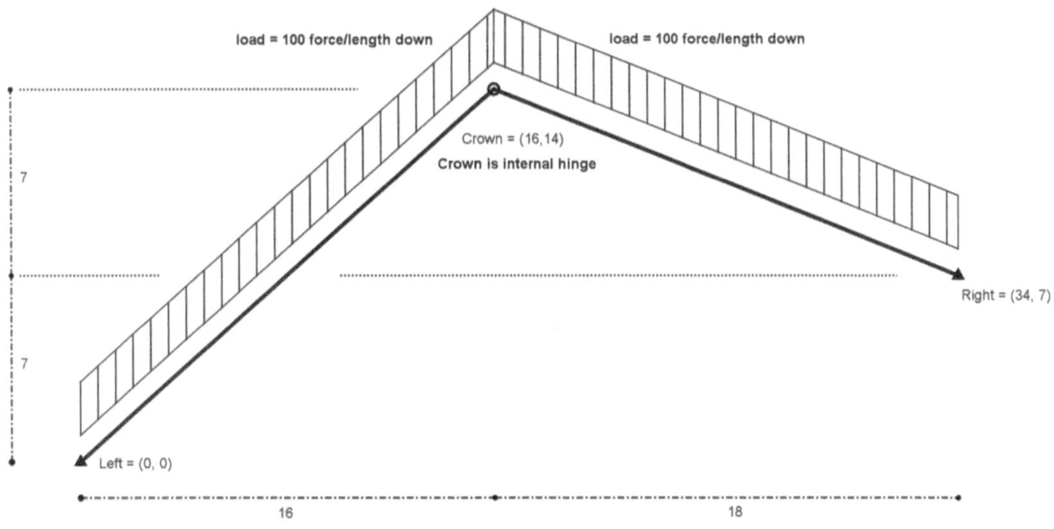

DOI: 10.4324/9781003180913-7

Even in the nineteenth century, structural designers readily understood the enormous power of the arch, and designers used them to span large distances, in many magnificent train stations around the world, in sports facilities, and for the support of long span bridge decks. The rise to span ratios, i.e. how high the crown of the arch is, compared to the width between supporting reactions, was largely driven by programmatic issues. For instance, how high must a train station roof be to adequately exhaust fumes? How high must the crown of a covered sports venue rise to accommodate spectator viewing, as well as the game itself? Another structural engineering issue that is affected by the rise to span ratio is the fact that "thrust" always exists, containing the arch. Thrust is the inward equilibrating reaction at the base of each leg of the arch. Even for downward only gravity loads, the legs would want to sprawl outwards if there was no such thrust containment. We call this the "giraffe on ice effect" which is immediately understandable in the reader's mind's eye. The giraffe on ice has only its own self-weight going downward, yet the spindly legs immediately want to spread out on the ice, whose surface cannot provide any thrust containment. Notice the triangular icons at the base of the arch legs in Figure 7.1. These carry horizontal force as well as vertical force. Thus, there are four unknown reactions in Figure 7.1. Yet the structure is still classified as determinate because of the internal hinge at the Crown. We can establish the magnitude of the reactions without taking into consideration anything about the stiffness of the materials used in the structure, this is the definition of "determinate". In general, a lower rise to span ratio means more and more thrust is developed at the equilibrating supports.

Brilliant designers have created astonishing arch structures for the past 150 years. Figure 7.2 shows the interior of the Galerie des Machines, engineered by Victor Contamin in 1889. Its span was an impressive 115 meters and the rise to the crown was 48 meters.

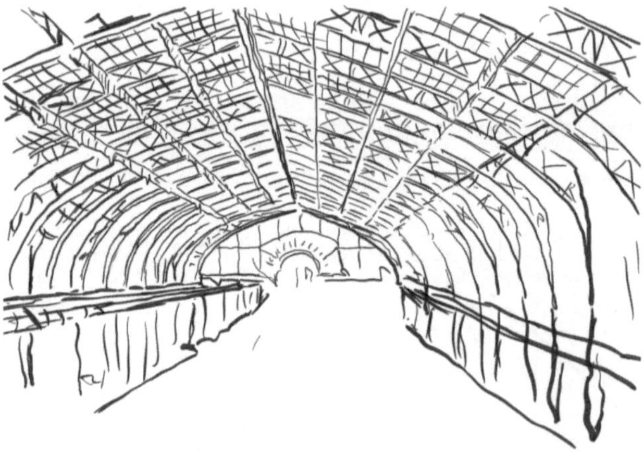

7.2
Victor Contamin, Galerie des Machines, Paris, 1889

In 1937, the prolific and important engineer Anton Tedesko designed the Philadelphia Skating Club, using two hinged, not three-hinged arches. Notice that the arches have significant depth at their crown. Would the reader be able to intuit how modifying that crown, such that it would actually be an internal hinge, would affect the thrust at the base? Tedesko analyzed this structure both as a two-hinged and as a three-hinged arch. It is still in terrific condition today, and is sketched in Figure 7.3.

7.3
Anton Tedesko, Philadelphia Skating Club

One of the greatest structural artists was Robert Maillart. His masterpiece, the Salginatobel Bridge continues to inspire us as a work of efficiency, economy, and elegance. Its graceful three-hinged arch is a determinate structure, whose form was determined by Maillart's deep understanding of how to control bending in reinforced concrete structures. Figure 7.4 shows the landmark Salginatobel Bridge from 1930.

7.4
Robert Maillart, Salginatobel Bridge, Switzerland, 1930

The secret used by Maillart in this bridge, and in other structures, is to create forms that are nearly "funicular." Funicular means the shape that a hanging chain would take under any loads. Maillart used graphic statics to create such shapes. Imagine a hanging chain, subjected to some loads, initially concave up, then frozen and subsequently inverted, so as to be concave down. Under the loads that deformed the chain originally, the new frozen concave down chain would not experience any bending at all. Essentially it would become a truss even though the load could be distributed along the length of its members. We mimic that form with a three-hinged arch, pinned at the lower left, hinged at the crown, pinned at the lower right. The closer the form is to the funicular, the less bending will be present under design loads. Figure 7.5 shows three such arches, each subject to identical point loads. Downward loads would cause a chain to form a segmented shape in the original, concave up shape. Inverting the segmented shape and making it concave down, forms the perfect funicular for the given loads. Each of the three arches shown with heavier lines in Figure 7.5 is slightly off of the funicular line, which is shown with lighter line weights, perhaps due to the programmatic reasons discussed at the beginning of this chapter. But as the offset becomes smaller and smaller, so does the bending get smaller and smaller. Figure 7.5 shows qualitative diagrams of the magnitude of bending for the three different structures subject to identical loads. Notice the wonderful design ideas generated by Figure 7.5! The shape of the arch could capture any one of these bending diagrams, creating a naturally graceful form. The last form in the lower right-hand side of Figure 7.5 is visually and technically linked to the form Maillart chose for the three-hinged arch that supports the deck of the Salginatobel Bridge. Look again at Figure 7.4 and the rightmost structure of Figure 7.5 and note the similarities.

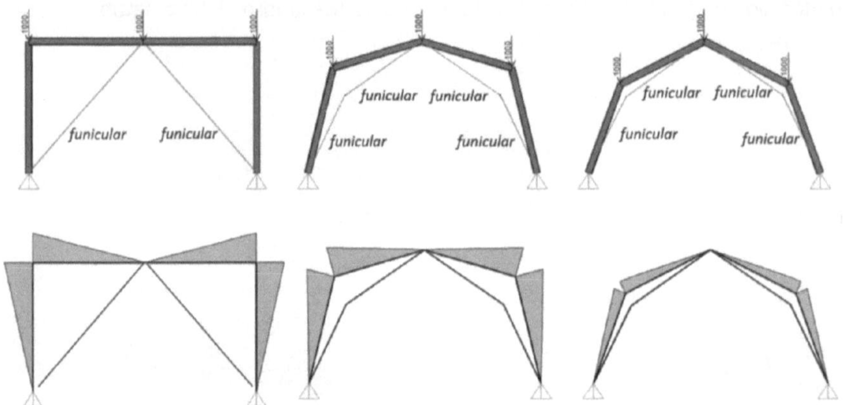

7.5
Distance from funicular line determines bending magnitude

Returning now to the problem originally shown in Figure 7.1, suppose we sought to find the magnitude of the equilibrating vertical reaction at the left support. Step 1 is to establish the magnitude and the location of concentrated point loads, which replace the uniformly distributed loads, one on either side of the kink at the crown. This is simple to do on a hand sketch. All that is needed to find the location is to determine the midpoint between the Left support and the Crown, and the midpoint between the Crown and the Right support. This is shown in Figure 7.6. To find the magnitude of the point load, multiply the force/length by the actual length it acts on. Refer back to Figure 7.1 to see the actual dimensions of the structure being studied.

7.6
Equivalent point load on either side of the Crown kink

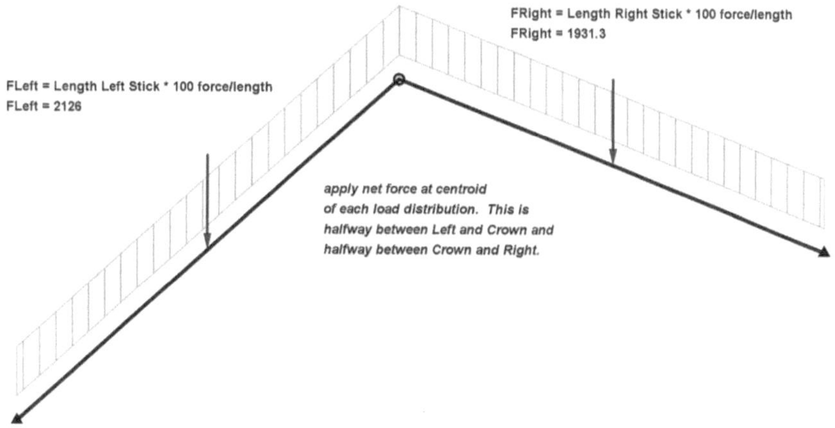

Since the vertical left reaction is sought, a vertical perturbation Δ is applied to the left support. Neither structural element changes length; thus, the New Crown is unambiguously found by intersecting the arcs of two circles. One circle is centered on the perturbed Left end, the other circle is centered on the unmoved right support. The left circular arc's radius is the original *chord length* of the left arch piece. This is a subtle but important point. It is the chord segment going from Left to Crown that swings through an arc length to establish the lofted Crown, called NewCrown in Figure 7.7. In this structure, the chord length is exactly the same as the original length of the member. But if the arch was shaped like the Galerie des Machines, the arch's curvilinear shape is irrelevant, only the radius from Left to Crown is needed, i.e. only the chord length. The same logic applies to the portion to the right of the crown. Its radius is the chord length from Crown to Right. This is shown in Figure 7.7. Once that New Crown is found, the perturbed shape is drawn. The point of load application is still midway between Left' and New Crown, and

7.7
Two arcs intersect to establish New Crown, lofts establish RLeftY for gross Δ

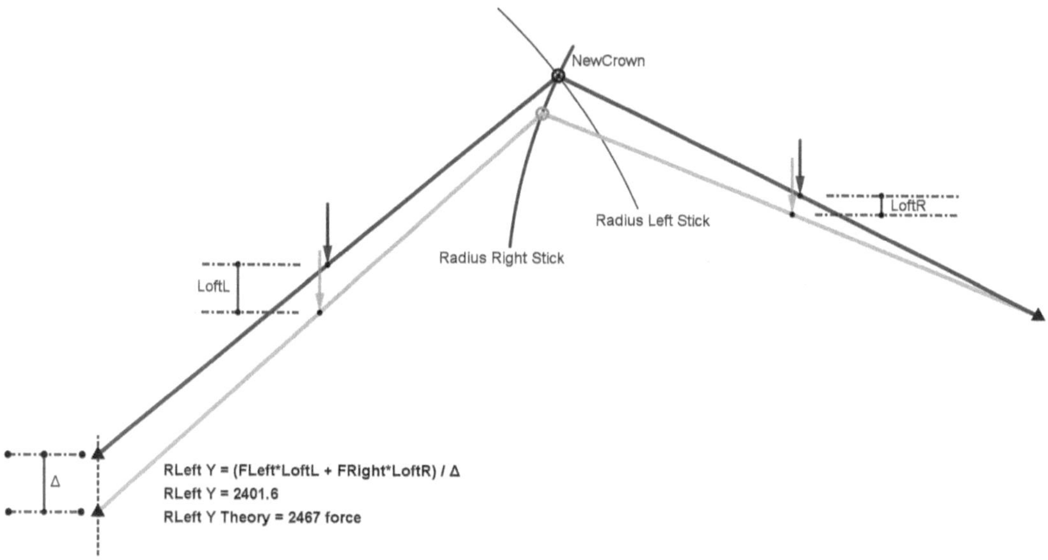

midway between New Crown and Right. Having the new points of load application is necessary to establish the lofts. Here the only information needed is the vertical loft since the applied loads are downward. Both loads cause negative work on the left-hand side of Equation 1.1 of Chapter 1. Then the Unknown, here the Left vertical equilibrating reaction, is immediately found.

Although the loads are only downward in this example, the members are inclined as they receive these downward loads. Thus, the Modern Müller-Breslau Method must be used. The answer will asymptotically approach the theoretical answer as the perturbation Δ gets smaller and smaller. This is shown in Figure 7.8.

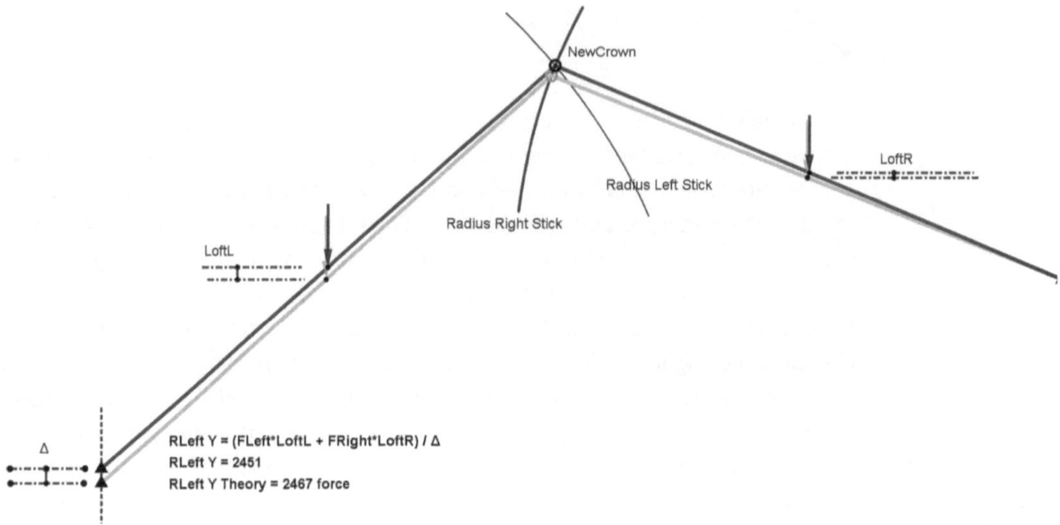

7.8
Two arcs intersect to establish New Crown, lofts establish RLeftY for small Δ

Of course, there was nothing special about starting with the left vertical equilibrating reaction. We could have started with an investigation of the left horizontal equilibrating reaction, here a thrust which contains the legs from sprawling outwards. If we assume an inwards perturbation, as shown in Figure 7.9, and if the Unknown Horizontal Force of Equation 1.1 of Chapter 1 is determined to be positive, then the thrust force is indeed inward. Figure 7.9 shows the solution for a large Δ, and Figure 7.10 shows the solution for a smaller Δ. Notice that again only vertical lofts are needed since we have only vertical externally applied forces FLeft and FRight.

Figure 7.11 is a study of a more complicated three-hinged arch, one that is similar to the arches in the Galerie des Machines. In Figure 7.11, there is a uniformly distributed lateral load, perhaps arising from wind. Notice that it is distributed uniformly along a vertical projection, as is traditional with Live Load. Notice also that a downward load is applied along the inclined length of only one portion of the arch, not along a horizontal projection. This is a requirement for Dead Load, it is always applied along the member length and is never projected. Such asymmetric loading is a source of problems in three-hinged arches. They tend to move or distort quite a bit when subjected to asymmetric loads. Figure

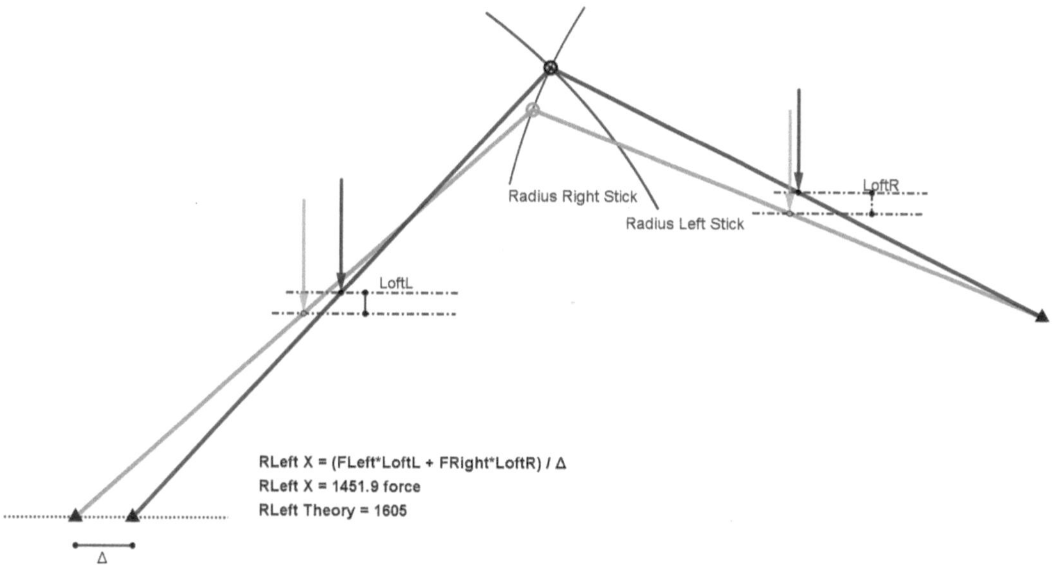

7.9
Two arcs intersect to establish New Crown, lofts establish RLeft X for large Δ

7.11 also suggest the enormous spans of the Galerie des Machines, if the units of length were meters.

The starting point and the ending point of each of the two distributed loads is fully known, so the centroid of each load is known, as the midpoint of each distribution. The equivalent concentrated forces are applied at the centroid, or mid-point in this case, of the distribution. This is shown in Figure 7.12.

Suppose the right containing thrust reaction force was sought. As such, the right reaction would be perturbed inward an amount Δ. As described previously, only chords from the supports to the crown are needed to swing arc segments that locate the New Crown. The intersection of those arcs establishes the unambiguous position of the New Crown. This is shown in Figure 7.13. What is not shown

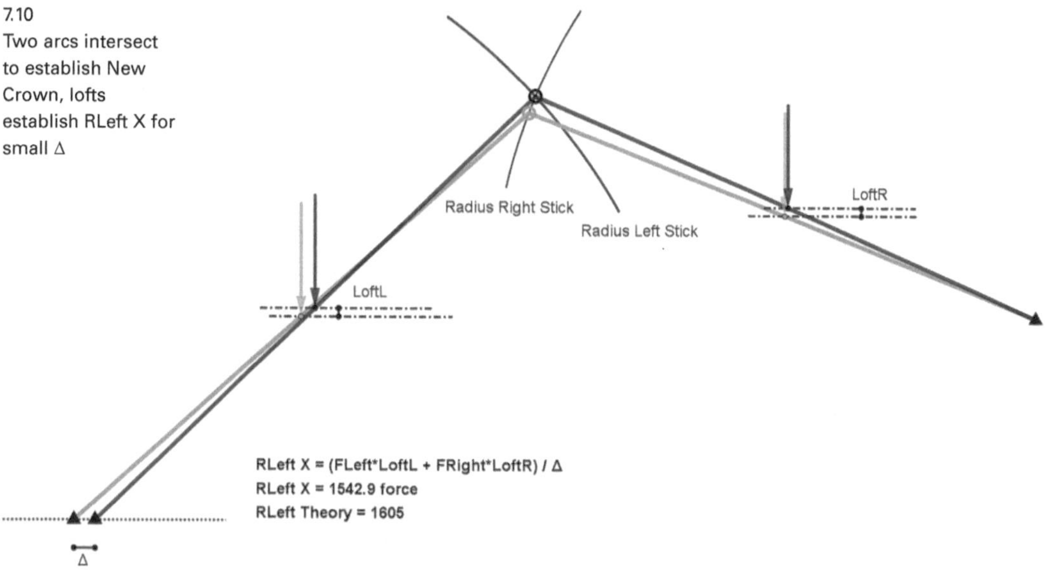

7.10
Two arcs intersect to establish New Crown, lofts establish RLeft X for small Δ

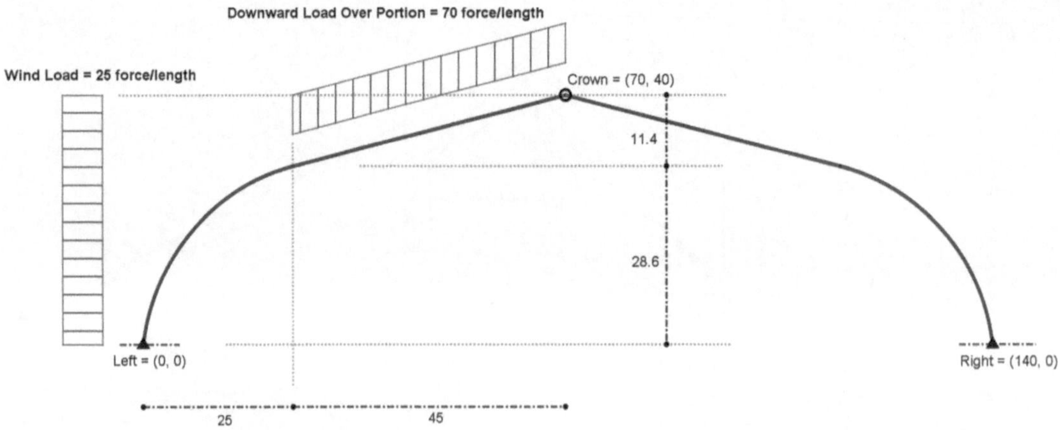

7.11 Galerie des Machines subjected to asymmetric loads

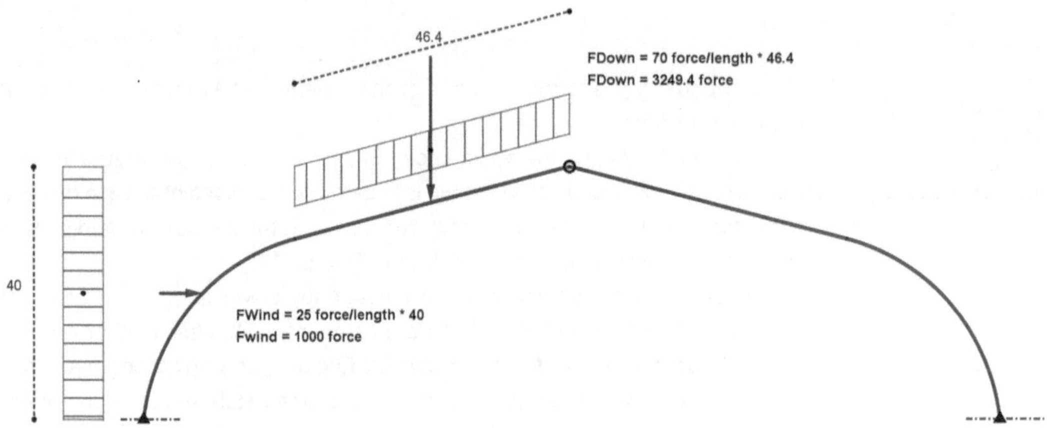

7.12 Galerie des Machines, two equivalent point loads

in Figure 7.13, is the determination of the new Load Points. They too are located simply by swinging chords from the original point of load application through the pinned support on the left. There are no loads on the right segment, but if there were, the new load location points would be established by radii that are chords to the perturbed right support. Even for the visibly large Δ of Figure 7.13, the answer is quite close to the theoretical value.

In another three-hinged arch shown in Figure 7.14, the structure rests on supports which vary in elevation. Perhaps the programmatic reason for this was the need for rows of stadium seats on the right-hand side. Such elevation changes pose no additional challenges to the Modern Müller-Breslau Method whatsoever. Notice in Figure 7.14 that the loads are concentrated at some known point along either side of the arch, with Cartesian Coordinates of the two load points shown.

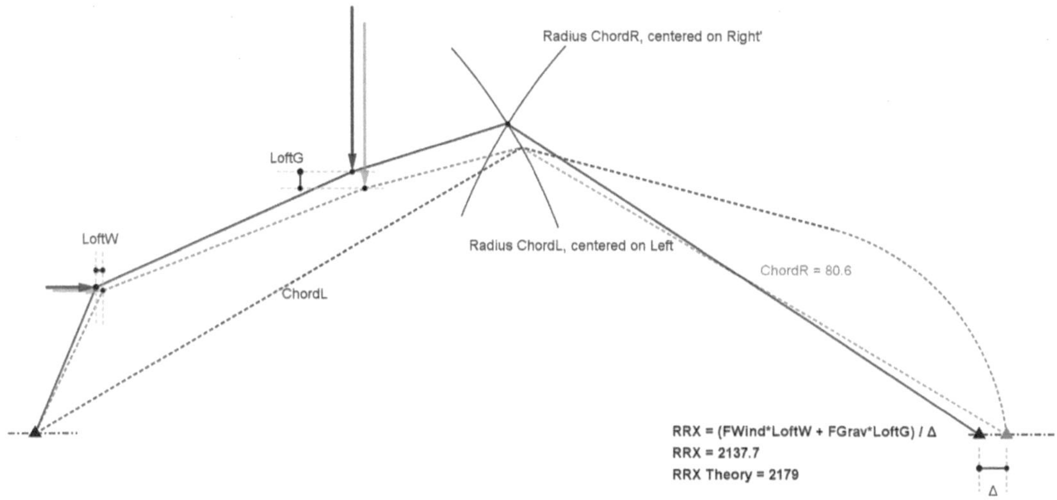

7.13 Establishing the right horizontal equilibrating force

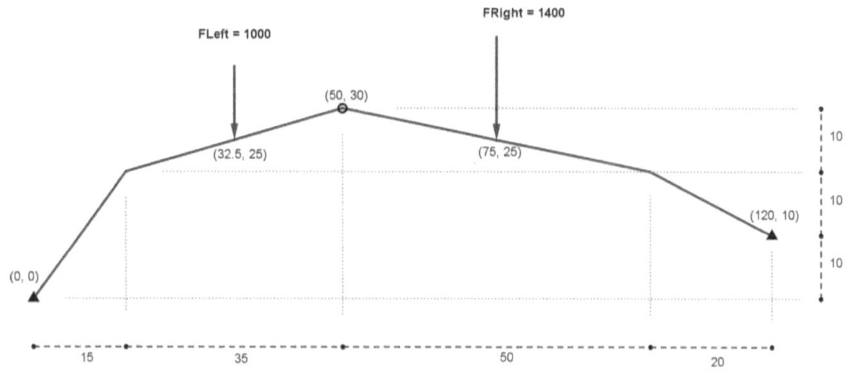

7.14
A large span three hinge arch with varying support heights

Suppose we sought the left horizontal thrust equilibrating reaction. As always, perturb the support in the direction of the desired unknown reaction. To unambiguously establish the location of the New Crown, create two separate arcs and find their intersection. The arc focused on the new Left' reaction has a radius of ChordL, the chord between the original Crown and the Left support. This radius has nothing to do with the loads, and nothing to do with the bend in the left arch. The arc focused on the unwavering Right support has a radius of ChordR, the chord between the original Crown and the Right support. That arc has nothing to do with the loads, and nothing to do with the bend in the right arch. Nevertheless, Figure 7.15 suggests a very quick method of establishing any point on the perturbed arch. Notice the dark point located at the bend on the right-hand arch segment. A chord from that point to the Right Support establishes the path which that point must follow. This technique must be used for any key points, such as the points where loads are applied or bend points which are moment carrying connections. They each follow their own arc, established by their specific chord length to the pinned support about which they rotate.

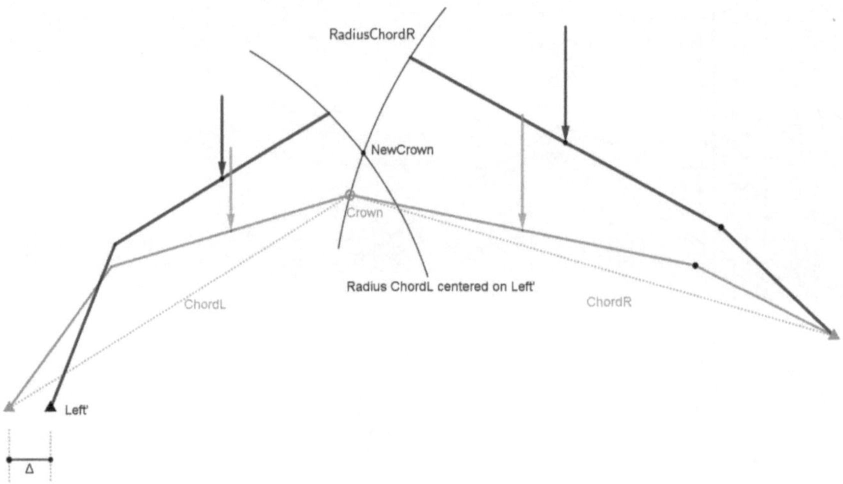

7.15
Chords establish the radius of the arc which any point swings through

Figure 7.16 shows a satisfactory geometry for the fairly large Δ. The term "satisfactory" here means several things:

- No segment of the arch changed length or shape.
- The angles at the original bends of segments remain the same in the lofted configuration.
- No boundary conditions are violated, no gaps between elements, no lifting off of required support conditions.

The second bullet point about the angles remaining constant is an absolutely critical concept that the novice reader may not have seen before. The *angle between two members* connected at a moment carrying joint in any arch or frame *remains constant*, even when the members themselves bend and curve. The novice reader should pause and visually check that the angles formed at each bend in the original configuration of Figure 7.16 are exactly the same as the angles in the perturbed configuration.

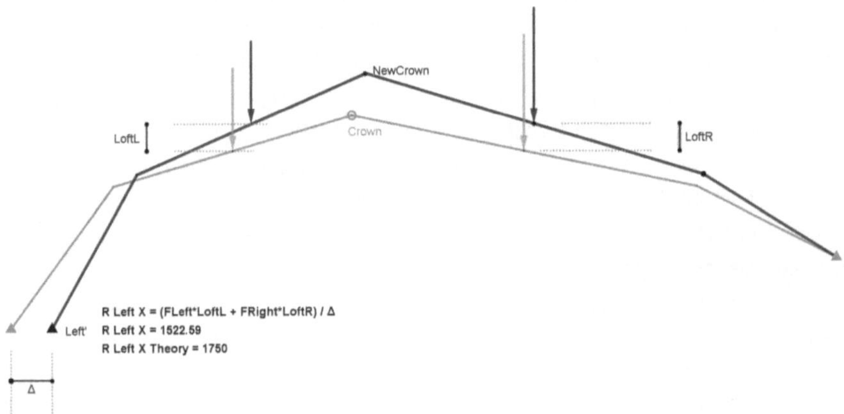

7.16
Chords establish the radius of the arc which any point swings through

As always, the theoretically exact answer is approached as the perturbation Δ gets small. This is demonstrated in Figure 7.17.

7.17
Final answer for thrust is approached

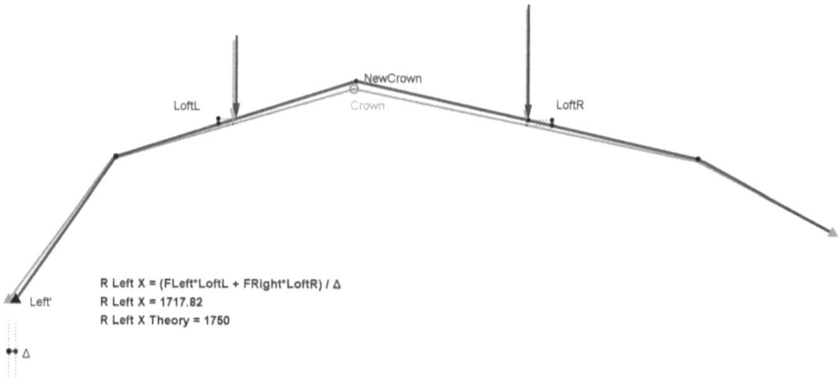

Consider now a more complicated arch structure, which could also be called a hinged frame, or a pitched frame. It is shown in elevation in Figure 7.18 and there are two main spaces formed by the structure. These spaces are known as "bays", they are in between the vertical columns. The roof is slanted, or pitched and notice the internal hinges at the crown of each bay. Also notice that the far right support is a roller, to ensure that the entire structure remains determinate. The structure is subjected to two lateral loads as shown in Figure 7.18.

7.18
Two-bay determinate structure

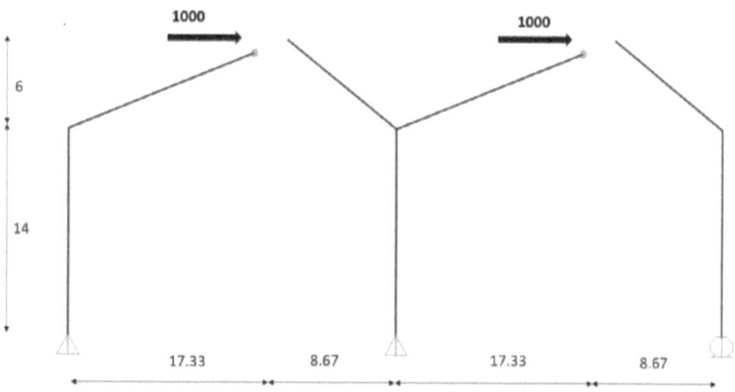

Suppose we sought the vertical reaction of the interior column. As always, perturb the column some amount Δ. In Figure 7.19, the perturbation Δ is the vertical shift of the interior support point. The other boundary conditions must be maintained! This means several things here. The interior support can have no horizontal perturbation as it is a pinned support. The right support can have no vertical perturbation as it is a roller support. But as such, the right support can indeed move sideways freely. And in all the problems concerning arches and frames, the original angles of moment carrying connection between elements must not change, i.e. all moment carrying connections of final angles are the same as the original angles. The pinned or internal hinged connections are allowed to change value freely. In

Figure 7.19, the reader should check that the angles formed at the "MoveMe" points have not changed in the perturbed configuration, whereas the angles at the internal hinge connections, symbolized by a black circle, did indeed change. As the structures get more and more complicated, the user will ensure that the boundary conditions are reasonably satisfied, with small errors. This will be accomplished by the manipulation of geometry in a parametric model. In a physical model, this would be very easy to do. These small errors will be described in the next chapter.

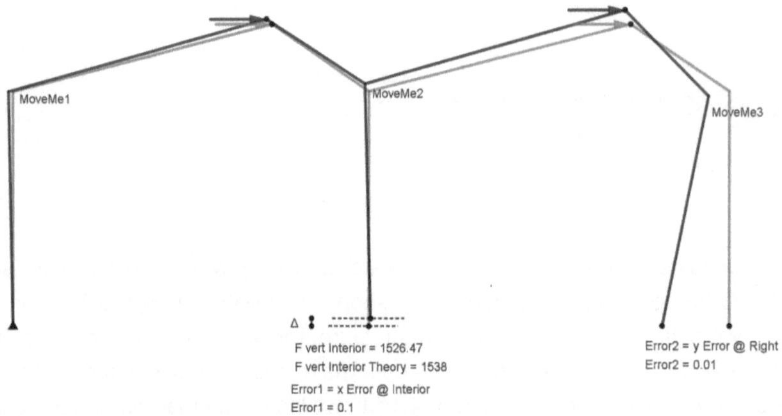

7.19
Two-bay determinate structure, solution for the vertical interior column reaction

The following structure setup is partially shown in Figure 7.20. There are three distinct bays with "pitched" or downward sloping roofs. In order to keep the form determinate, care has to be taken to ensure that only certain joints transfer moment, these are shown as having no gap between connected elements. Other connections are internal hinges, which do not transfer moment, these are shown with a gap between connected elements. Note that the left support is a roller which is free to translate only horizontally, but cannot move up or down at all. Before we apply loads and specific dimensions to the structure of Figure 7.20, imagine that we sought the right interior column's equilibrating horizontal or thrust reaction to any loads, i.e. what is the horizontal force in the third column pinned base?

7.20
Three-bay determinate structure, seeking the third column horizontal reaction

The Modern Müller-Breslau Method always looks at the deformed structure due to a specific perturbation of the unknown, never ever due to deformations caused by loads. So regardless of the loads, the perturbed structure looks like the one shown in Figure 7.21, when seeking a horizontal reaction at the third column. Here, a large leftward Δ of the third support is shown, this assumes that the reaction is leftward. If the answer for a particular set of applied loads is negative, then the reaction would be in the opposite direction of the assumed Δ.

7.21
Three-bay determinate structure, perturbing the third column base horizontally

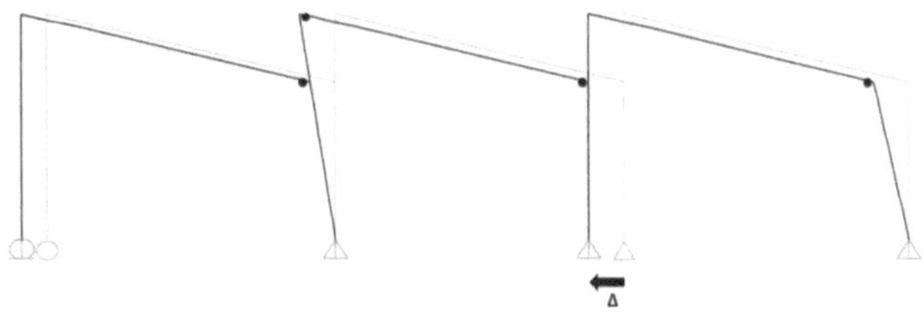

In Figure 7.22, we seek the third column's vertical base reaction. Thus, a large upward Δ is shown for the third column base. If the answer for a particular set of loads is negative for such a perturbation, then the reaction would be downward, not upward.

7.22
Three-bay determinate structure, perturbing the third column base vertically

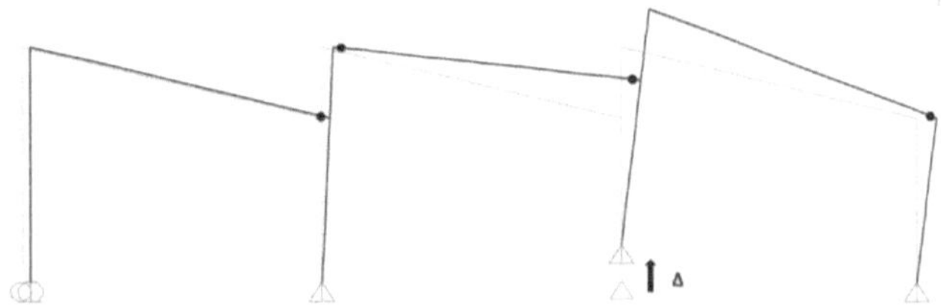

The reader should notice in Figures 7.21 and 7.22 that the moment carrying joints, shown with no gap between elements, maintain the original angle between connected members, through the lofted configuration. But members which are hinged together, shown with black circles, are free to distort that angle between hinge-connected members, which is the definition of an internal hinge. Suppose this three-bay structure had the dimensions and loads shown in Figure 7.23. We seek the vertical reaction at the third column base.

7.23
Three-bay determinate structure, seeking the third column base vertical reaction

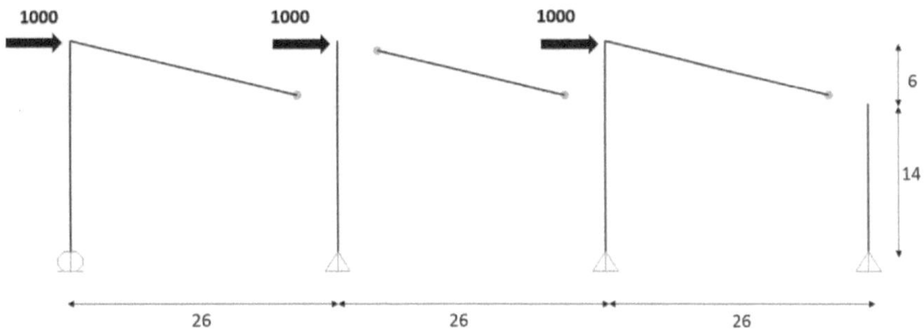

Figure 7.24 shows a large vertical perturbation Δ at the Unknown, yet the other boundary conditions are nearly satisfied, with small errors that can be adjusted by the user. The coarse estimate of the vertical reaction force is shown in Figure 7.24. The reader may be surprised that the equilibrating reaction is downward, i.e. it pulls the structure back to the ground.

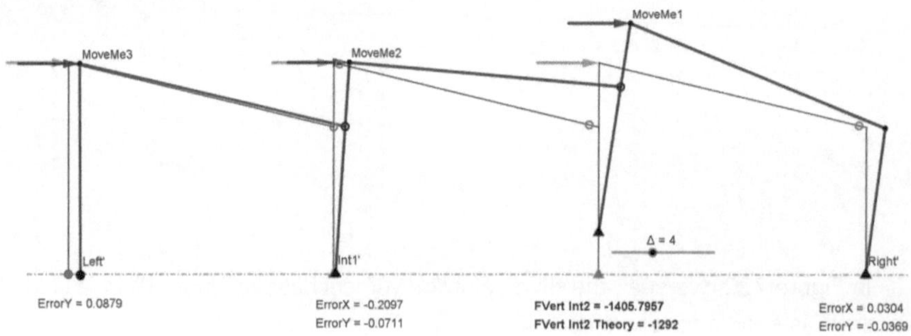

7.24
Three-bay determinate structure, an estimate of the third column base vertical reaction

For smaller perturbations, with boundary conditions that are nearly satisfied, the answer approaches the theoretical value. This is shown in Figure 7.25.

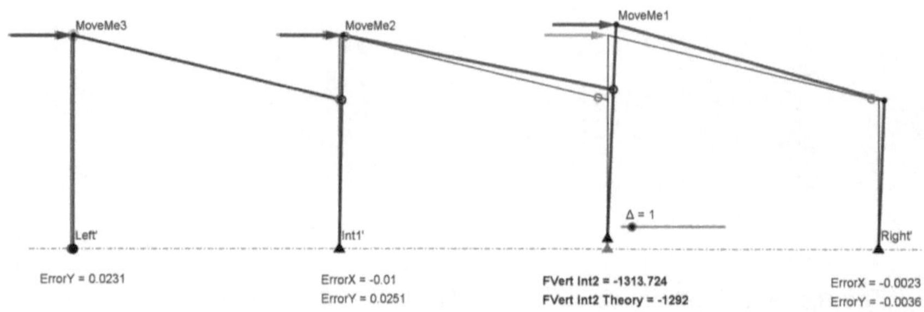

7.25
Three-bay determinate structure, converging on the third column base vertical reaction

The following structure has two bays, and each bay has a "pitched" or steeply sloped roof. Notice the internal hinges, symbolized by the gaps in a connection in Figure 7.26. Also notice that the far right support is a roller, which is free to move horizontally, but cannot move up or down. Before any loads or dimensions are applied to the structure of Figure 7.26, imagine the perturbed shape if the vertical reaction of the interior support was sought. The reader should sketch out the perturbed shape, which is always independent of any loads that are applied. Also, the angle between the two members connected by the sole moment carrying connection must remain the same throughout the perturbation. All the other angles will change throughout the perturbation.

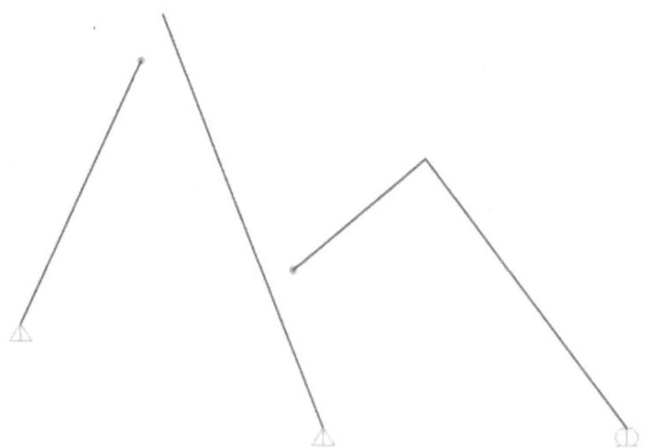

7.26
Two-bay determinate structure, seeking the interior column base vertical reaction

Figure 7.27 shows a large vertical perturbation at the interior column's base.

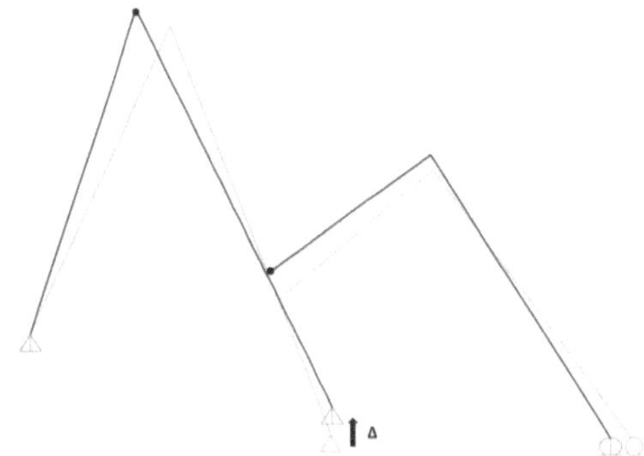

7.27
Two-bay determinate structure, perturbing the interior column base vertically

Applying the Modern Müller-Breslau Method to this two-bay structure with the following loads and nodal coordinates, we seek the interior pinned support's vertical equilibrating reaction. Figure 7.28 shows the details.

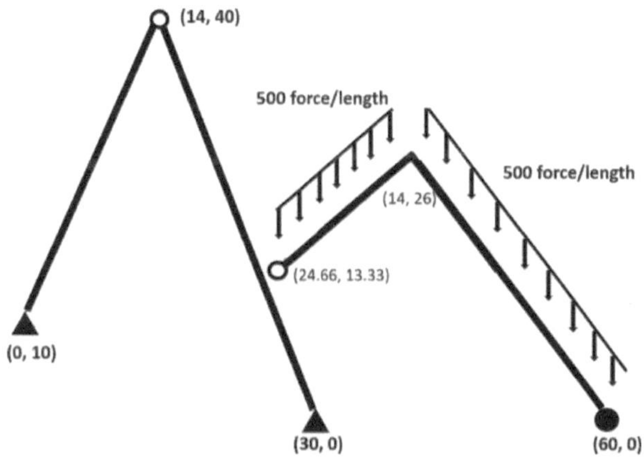

7.28
Seeking the interior vertical equilibrating reaction

Figure 7.29 shows the solution for a large perturbation and Figure 7.30 shows the solution for a small perturbation. The reader is encouraged to verify that the sole moment carrying connection maintains its original angle throughout the perturbation.

Returning to the arch originally shown in Figure 7.1 of this chapter, suppose we sought the moment midway along the left arch member. We perturb the single element, here the left arch member by inserting a hinge, or removing the capability of the element to transfer moment at the cut. That single element is stretched on either side of the cut, but no other elements are elongated. This requires the right element to remain unmoved during the perturbation. This is shown in Figure 7.31.

The Modern Müller-Breslau Method immediately presents an insight that might be surprising to the novice reader. That is the fact that the load on the right side of the arch has absolutely no influence at all on the moment being studied on the left side of the arch. This is visually obvious because the perturbed configuration

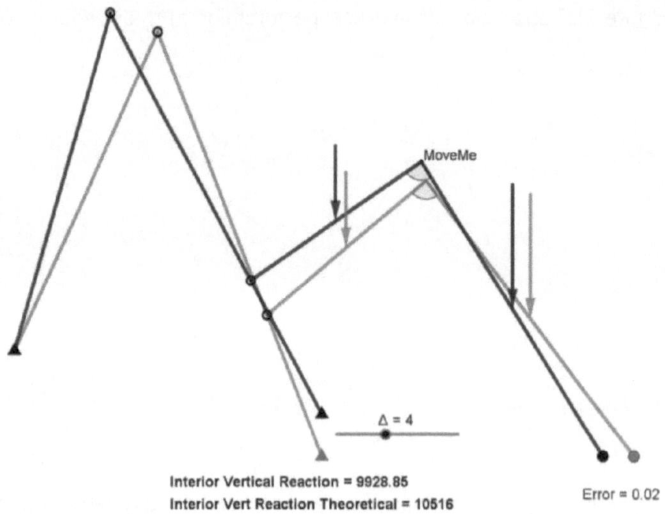

7.29
Interior vertical equilibrating reaction for large Δ

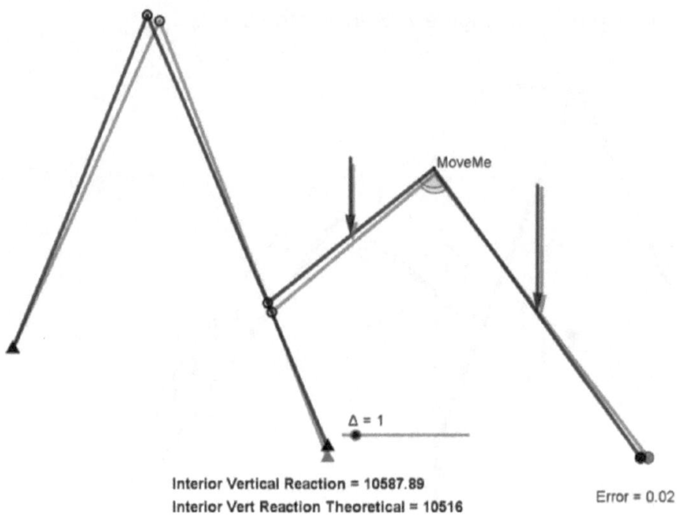

7.30
Interior vertical equilibrating reaction for small Δ

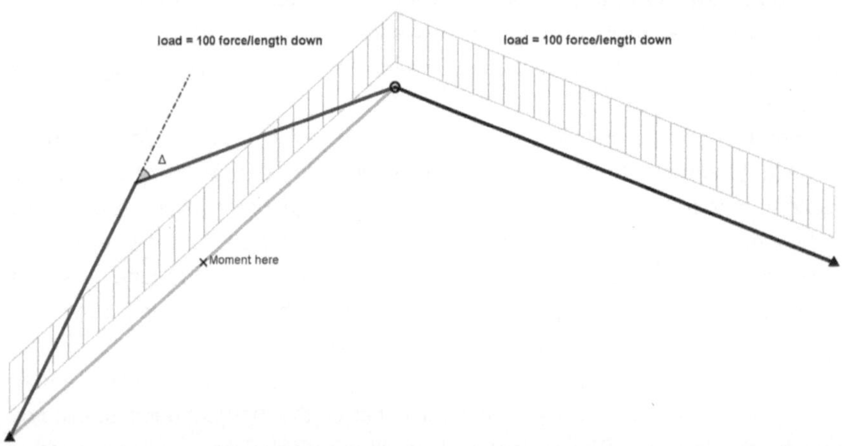

7.31
Seeking moment at midspan of the left element

has zero effect on the right side of the arch. The technical reason for this is linked to the mini-study shown previously in Figure 7.5. For loads on the left side of the arch, bending occurs on the left side of the arch. But for loads on the right side of the arch, no bending occurs on the left side of the arch, because the forces from the right side of the arch are solely transferred to the left side of the arch through the crown, not through bending of the straight left element. In other words, the load on the right side becomes a nodal point load at the crown in its affect on the left side. If however, the left side was bent as on the far right of Figure 7.5, then bending would indeed occur on that left side due to loads applied on the right side. This is so because the bent element would be slightly off of the funicular line required by concentrated transferring forces in the crown, or in other words, a funicular for transferring point loads at the crown.

Since a kink has developed under the distributed load on the left element, an equivalent concentrated force on either side of the kink must be applied. Then the lofts of Part1 of the loads left of the kink and of Part2 of the loads, right of the kink are found. This is shown in Figure 7.32. Notice that since the distributed load is downward, we seek only vertical lofts.

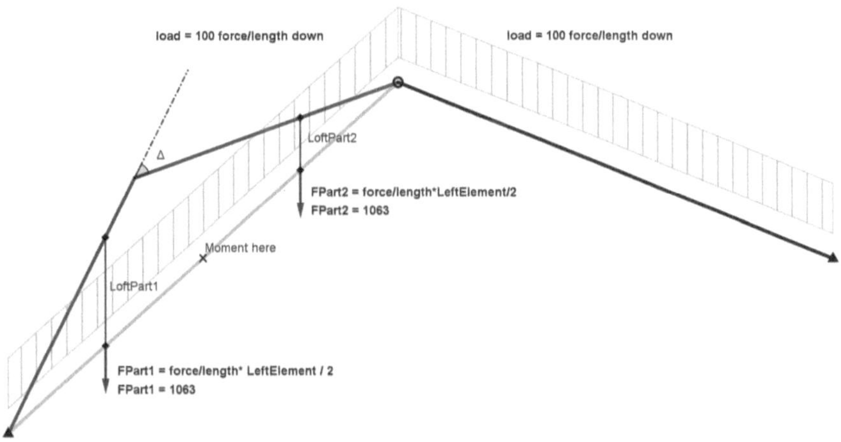

7.32
Loads on either side of the kink must be established

As the perturbation Δ gets small, the moment approaches the theoretical value as shown in Figure 7.33.

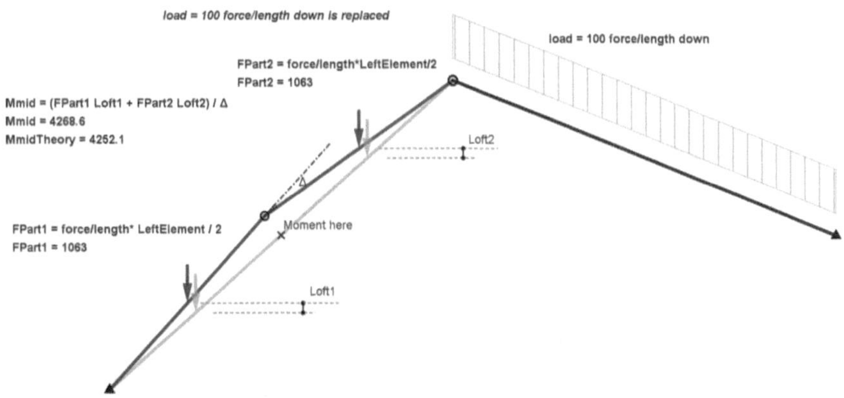

7.33
Solving for a moment halfway up left arch element

Consider finally, a three-hinged arch which has normal loads, i.e. loads that are perpendicular to (normal to) the left member, and gravity loads on the right member. A typical situation is shown in Figure 7.34.

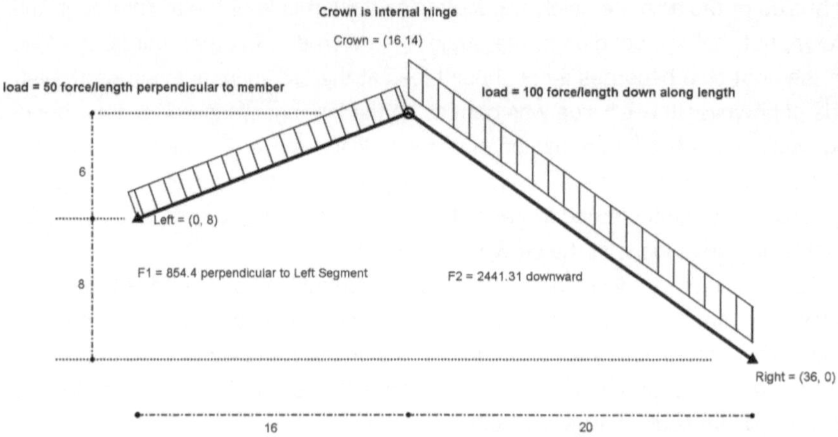

7.34
Three-hinged arch with a normal load and a downward load

If we seek the moment at midspan in the left element, we perturb that element by "cracking" it, to find the unknown moment. The loads must be broken up proportionally on either side of the kink formed by the crack, here each load would be one half of the total normal load. But the lofts must be established in the direction of the load, as required by Equation 1.1 of Chapter 1. A qualitative investigation is shown in a hand sketch in Figure 7.35. Note in Figure 7.35 that the downward gravity loads induce no bending at all in the left-hand member. This may surprise the novice reader!

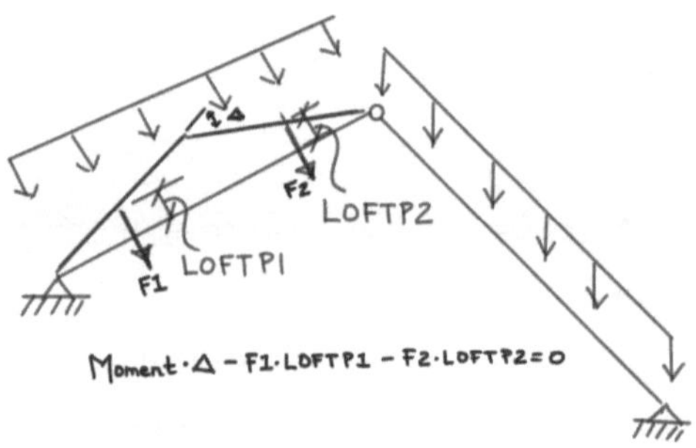

7.35
Qualitatively finding M midspan on left element

ARCH ÉTUDES

The following studies or études should be sketched by hand.

Étude 7.1) A three-hinged arch is shown in Figure 7.36. It will be used for several études. Sketch it and calculate the total load on the left member and the total load on the right member.

Answer: Force Left = 854 units of force, perpendicular to member Force Right = 2,441 units of force downward.

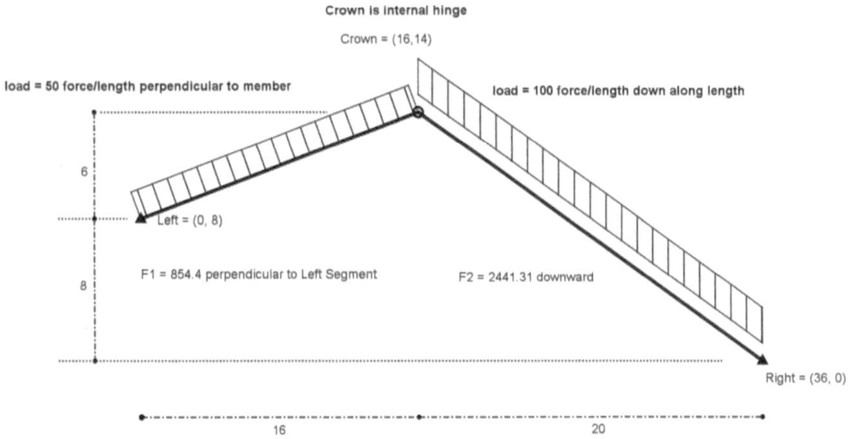

7.36
Étude 7.1 setup

Étude 7.2) Sketch out a perturbed shape for a large Δ, perhaps 5 units of length, if we are seeking the left vertical equilibrating reaction of the structure in Figure 7.36. Take care to ensure that the loads remain at their original orientations in the perturbed shape. Do not calculate the actual left vertical reaction, keep everything qualitative. A hand sketch solution is shown in Figure 7.37.

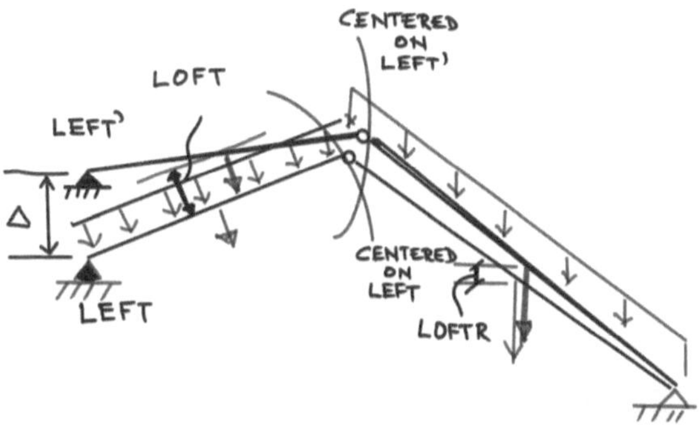

7.37
Étude 7.2 solution

Étude 7.3) Repeat étude 7.3 but now calculate the left vertical reaction for this large perturbation quantitatively. A hand sketched solution is shown in Figure 7.38.

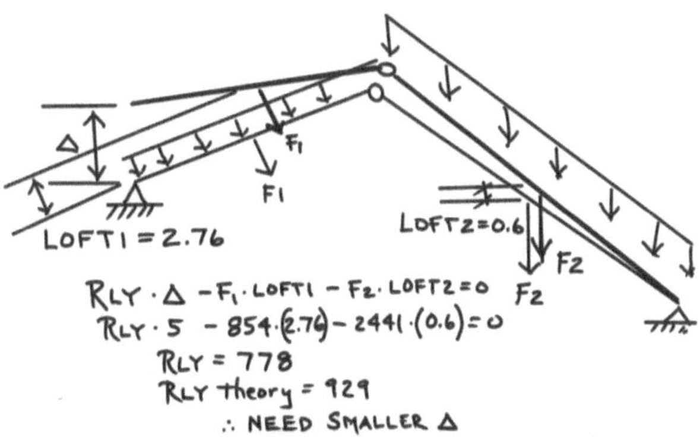

7.38
Étude 7.3 solution

Étude 7.4) Use a large perturbation to find the internal moment at midspan of the right member shown in Figure 7.36. Be sure to keep the load on the right part vertical, and be sure to place proportional loads on either side of the crack, here ½ and ½ of the total load going down on the right. A hand sketched solution is shown in Figure 7.39.

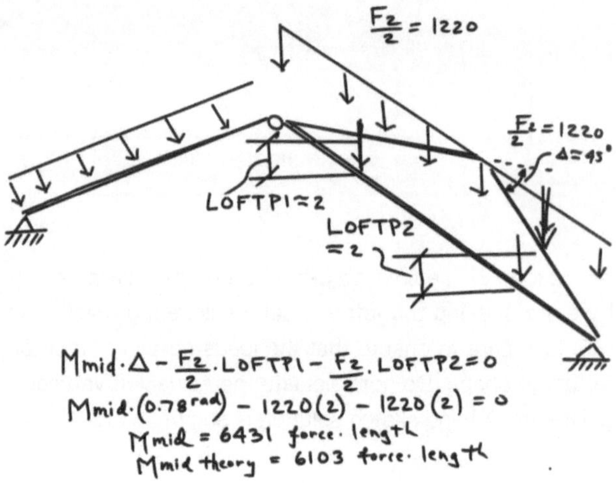

7.39
Étude 7.4 solution

Étude 7.5) A three hinged arch is shown in Figure 7.40. Sketch out by hand the unperturbed geometry. The base of each column is fully fixed. Notice the three internal hinges and that the roof load is downward. Calculate the total downward load.

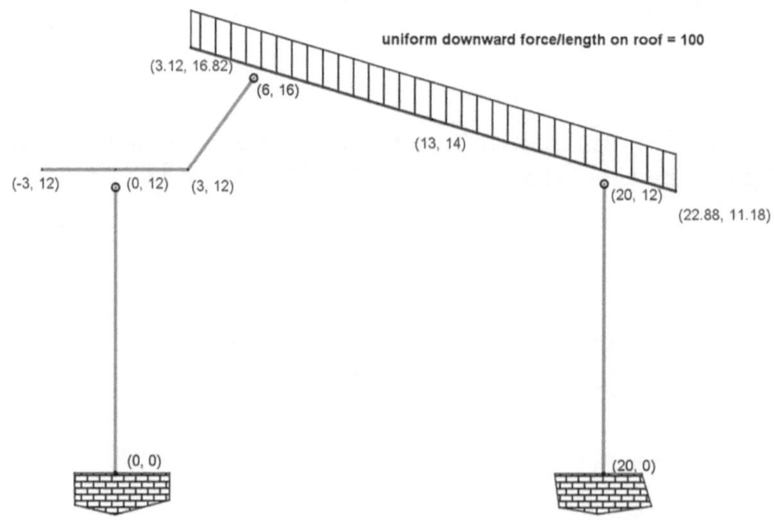

7.40
Étude 7.5 solution

(Ans: F roof total = 2,056 units of force)

Étude 7.6) For the structure of Figure 7.40, sketch out by hand the perturbed shape for a large Δ of 2 units of length applied to the vertical axial reaction at the left base. Note that the structure is fixed at its left and right base points; yet it is still determinate because of the three internal hinges. Estimate as best as you can how much the free tip at the cantilevered edge of the roof on the left side is lofted. A hand sketched solution is shown in Figure 7.41.

7.41
Étude 7.6 solution

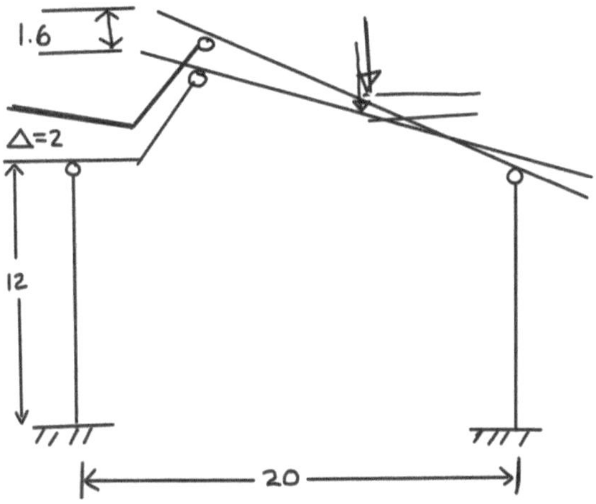

Étude 7.7) If the free left roof tip is at 1.68 units of displacement for a Δ of 2 in Figure 7.41, find the equilibrating axial force in the left column. A hand sketched solution is shown in Figure 7.42.

7.42
Étude 7.7 solution

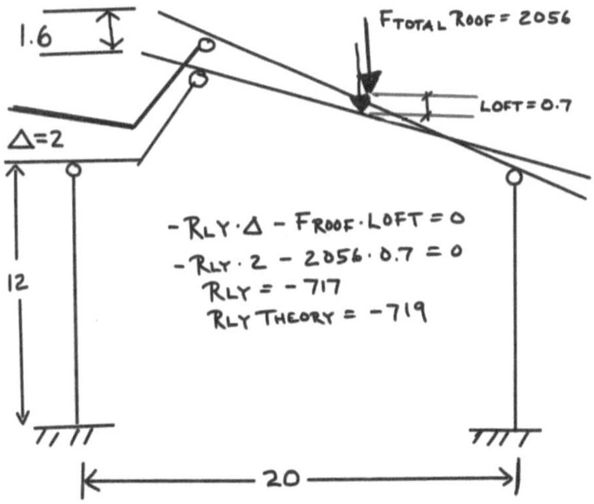

Étude 7.8) For a large Δ at the midspan of the roof for the structure of Figure 7.40, which is subjected to a uniform load of 100 force/length downward, find the moment at the midspan of the roof. A hand sketched solution is shown in Figure 7.43.

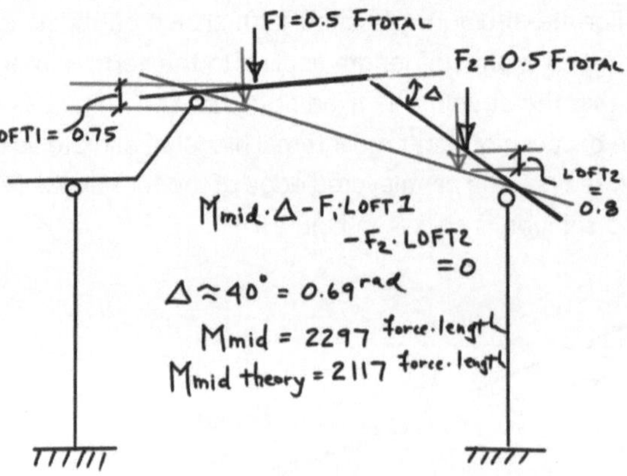

7.43
Étude 7.8 solution

8 GeoGebra for Arches

The basic steps of any parametric program, in any language, will be custom designed by the author of the program, with personal touches that are sensible and logical to the creator. Yet, some helpful guidance and hints may make the task less daunting for readers not accustomed to writing such scripts. A few of the arches of the previous chapter will be broken down into easy to follow steps. Readers may well find more logical or more efficient ways of obtaining these solutions.

Several fundamental rules governing all the solutions for *external equilibrating reactions* at supports of a determinate arch, whether they are vertical or horizontal are that:

- Individual elements of an arch do not bend.
- Individual elements of an arch do not change the length.
- Distributed loads are applied at the centroid of the distribution, on either side of any kink.
- Moment carrying connections never change angles throughout the perturbation.
- Internal hinge connections freely change angles throughout the perturbation.
- The user may have to manually adjust supports to closely, but not perfectly match support boundary conditions.

The simplest three-hinged arch is one that is composed of two straight elements. Notice an immediate quandary, one can specify the lengths of the two elements, or one can specify the locations of the three hinges, but one cannot specify both of these arbitrarily. It is more practical to specify the lengths of individual elements of the arch, as that as what a builder would bring to a construction site. Suppose there were two elements, one is 15 units of length long, and the second is 20 units of length long. Furthermore, assume that the construction grid set up Cartesian Coordinates of (0, 8) for the base support of the first element. Points in GeoGebra always begin with a capital letter, thus inputting the following short lines of code would be a reasonable starting point. Recall that the Input Bar at the bottom of the page is where all commands are entered.

- L1 = 15
- L2 = 20
- Left = (0,8)

DOI: 10.4324/9781003180913-8

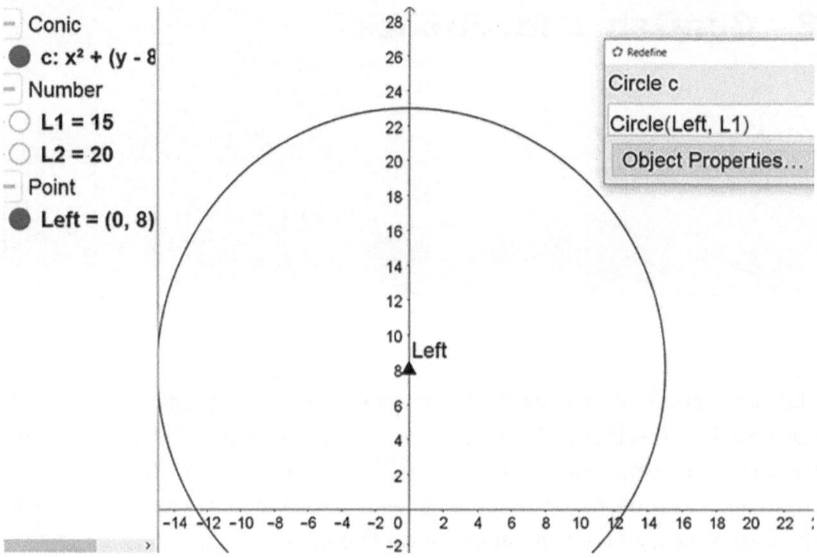

8.1
Initial setup for a three-hinged arch

In this first example, the coordinates of the Right support and of the Crown meeting point are not yet known.

A circle of radius L1, centered on the Left support can be drawn. This is shown in Figure 8.1.

Although the final Crown must be somewhere on this circle, do not place a point that is assumed to be the Crown on the circle just yet. The Crown will be found from the intersection of two circles, whose radii are the arch elements' lengths. In this first example, the coordinates of the support for the second arch element are user defined but are not 100% arbitrary, because if the Right support is too far away from the Left support, the circles will never intersect. The user should either type in Right = (#*x*, #*y*) where #*x* and #*y* are Cartesian Coordinates of the Right point, or alternatively, a Point can simply be dropped on the canvas space.

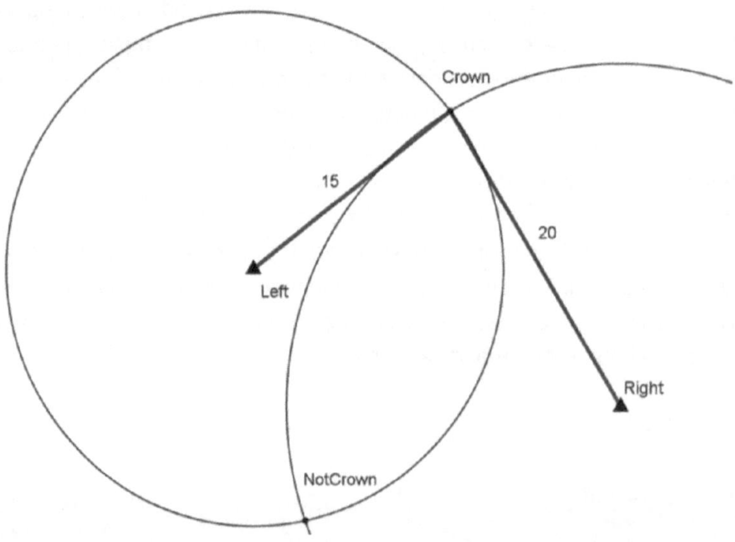

8.2
True Crown is found from the intersection of two circles or radii L1 and L2

A second circle centered on Right, of radius L2 is drawn. The intersection of the two circles is found. Note that there are two intersections, but only one is practical from a building point of view, the second intersection point, labeled as "NotCrown" would be below the ground. This is shown in Figure 8.2.

While there are many different ways of perturbing any support, either vertically or horizontally, a particularly quick way of doing this is to input the perturbation value Δ either on a slider or as a single-valued entity in the Algebra Window. The Slider icon is found on the second to last drop down menu on the top of the workspace. Reasonable limits for Δ might be 0–10 for example. Then, in the Input Bar, creating a perturbed point can be achieved by defining Left′ = Left + (0,Δ) if one was seeking the vertical equilibrating reaction at the Left support. Then the Crown′, which is the lofted Crown, is found by the exact same process as before, i.e. a circle centered on the perturbed Left′ with radius L1, and a circle centered on the unmoved Right, of radius L2. The intersection of these two circles is Crown′, with the spurious intersection below the ground ignored. Notice that we never delete unnecessary items, deleting the second intersection accidentally deletes the needed Crown′ as well. This geometric construction is shown in Figure 8.3.

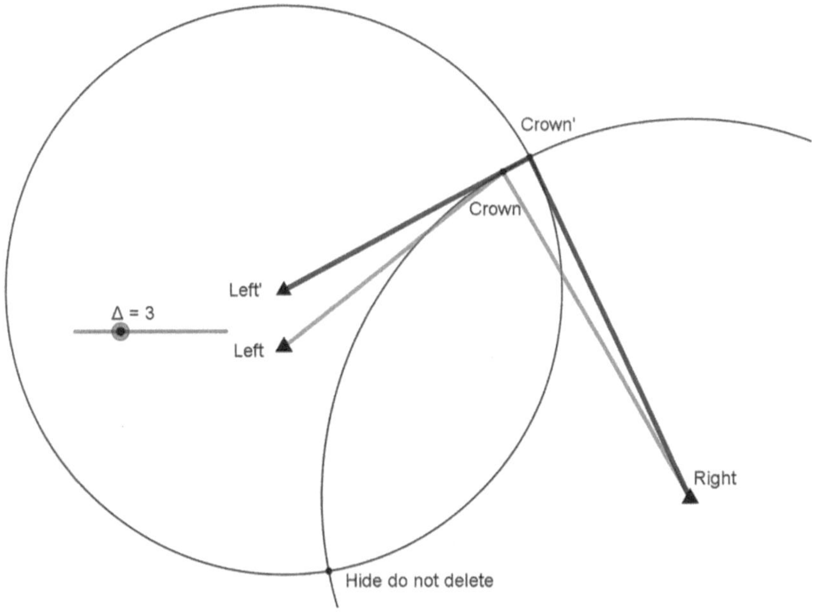

8.3
Lofted Crown is found from the intersection of two circles or radii L1 and L2

If the load is applied at the midpoint of an element, it is simple to find the midpoint of the lofted element. But suppose the load was applied somewhere else along the element, not at an easy to define spots such as at the midpoint, or a point a third or a quarter of the length of the element. This is a very important idea for the reader to grasp, as it is applicable to a wide variety of uses in these geometric constructions. For example, suppose a load was applied at some known point along the length of the left element in this three-hinged arch. Where would the load be located on the lofted element? This question is posed in Figure 8.4.

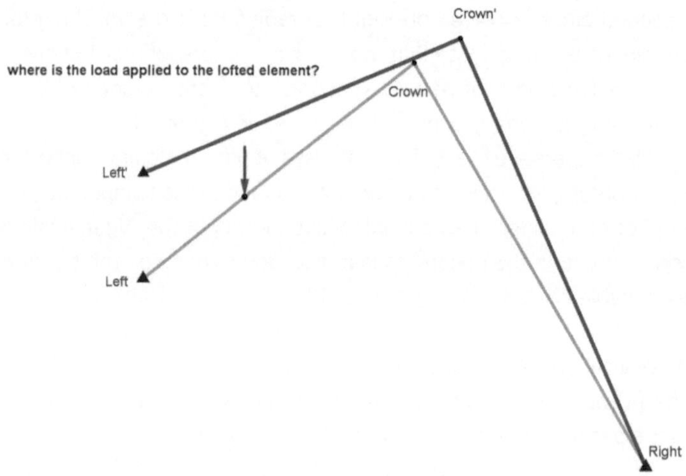

8.4
Finding where the load is applied along the lofted element

Clearly, there are many of establishing that point, but a very quick and easy way of doing so is to:

- Hide, do not delete the left element.
- Establish a segment from Left to the point of load application.
- Create a circle whose radius is that segment length, centered on Left'. Alternatively, and perhaps more simply, one can draw a Circle with Center Through Point, and snap to the point of load application.
- Intersect that circle with the lofted element.

An extremely handy feature of GeoGebra is the fact that the length of the segment can be used by simply using the name of the segment itself! In many programs, one would have to establish Length(segment) to find the scalar length. Here, we can simply create a circle whose radius is "segment" where "segment" is the name of the segment. Figure 8.5 demonstrates this important idea:

There are at least two different ways of establishing the loft of the load:

- LoftLeft = *y*(LoadPt') – *y*(LoadPt)
- LoftLeft is the name of a vertical segment whose ends are defined by parallel horizontal lines passing through LoadPt' and LoadPt.

Figure 8.6 shows the calculation of the loft taken by the load.

The final step is to write one line of code that solves for the Unknown being sought, from Equation 1.1 of Chapter 1. This is placed in the Input Bar, as are all commands.

$$Unknown = \frac{ForceLeft \cdot LoftLeft}{\Delta} \qquad \text{(Equation 8.1)}$$

As the perturbation Δ gets smaller and smaller, the Unknown answer will approach the theoretically correct value.

For arches that are curved, or that have distinct straight segments separated by bends, the technique is exactly the same as that which was shown in Figure 8.5, namely to establish chords from key points and to swing that chord through a circular

8.5
Locating the load along the lofted element

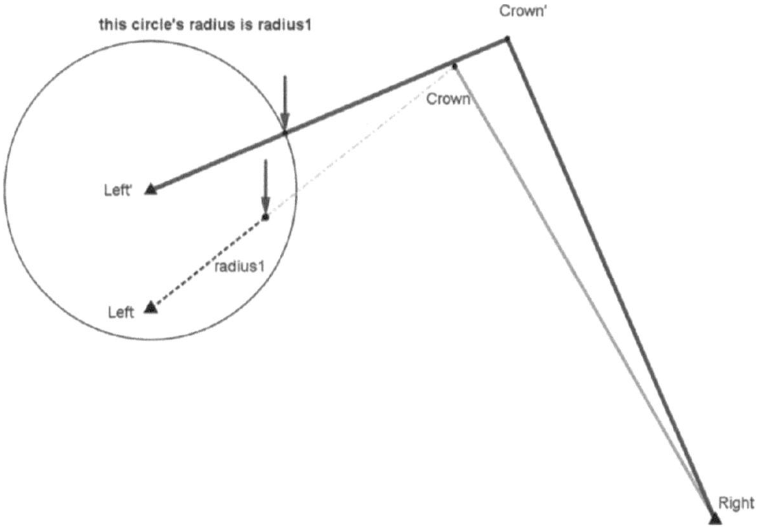

8.6
Establishing loft of the load

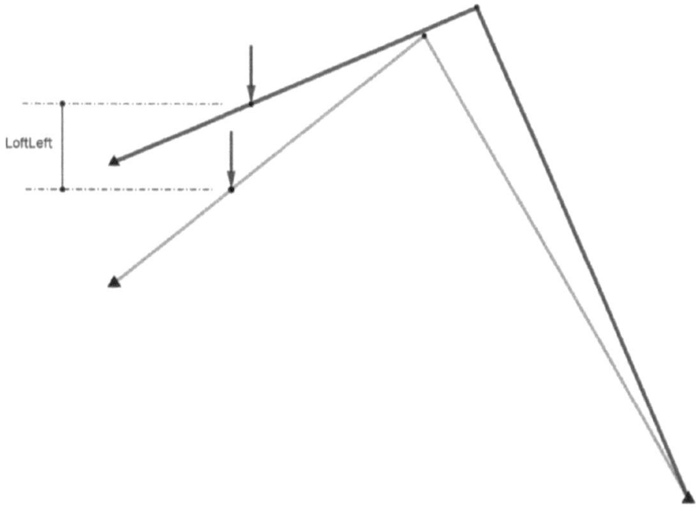

path to find the perturbed configuration. For example, to locate the bend between the first two segments in the perturbed shape of Figure 8.7, notice that the original bend lies on a circular path of radius Element1, centered on the Left support. Thus, the perturbed bend must lie on a circular path of radius Element1, centered now on the perturbed Left' support. This much is hopefully obvious and easy to see for the reader, but exactly where on this circular path the point lies is still a mystery!

There are several steps that ensue, to establish where this bend is located. The first step was shown in Figure 8.7, in that we certainly know that the bend is somewhere on the circular path whose radius is the length of the first element. Yet ultimately, there are still two issues that absolutely must be resolved to establish the perturbed geometry.

- Moment carrying connections between elements never change angles in the perturbed configuration.
- Chord lengths establish radii that are swung through circular paths.

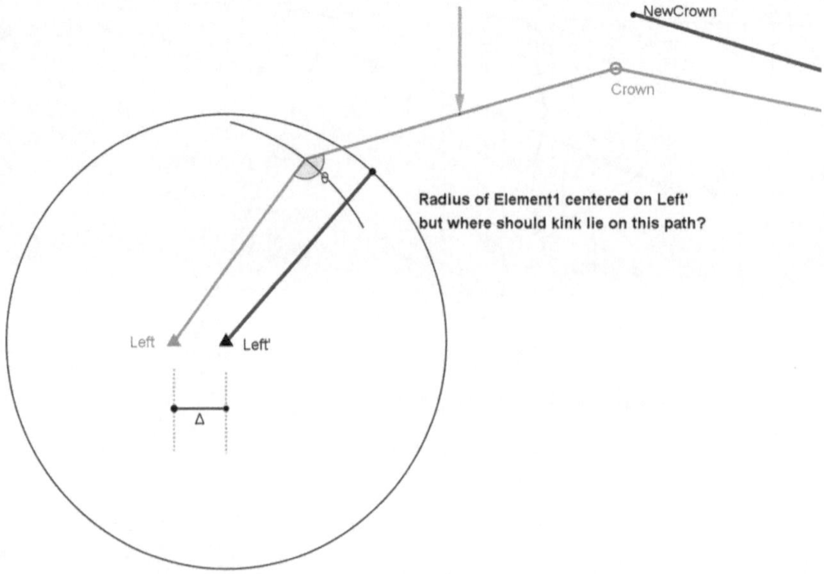

8.7
Step one of creating a perturbed shape

Figure 8.8 shows how the angle θ is maintained through the perturbation. GeoGebra has two handy angle features. The first solely establishes an angle through three points. Figure 8.8 shows this first feature, as the angle θ is defined by three points, Left, Bend, and Crown. The second angle drop-down icon re-creates an established angle through a leg point and a vertex. Figure 8.8 also shows this feature, in that GeoGebra creates a line that specifies the previously established angle.

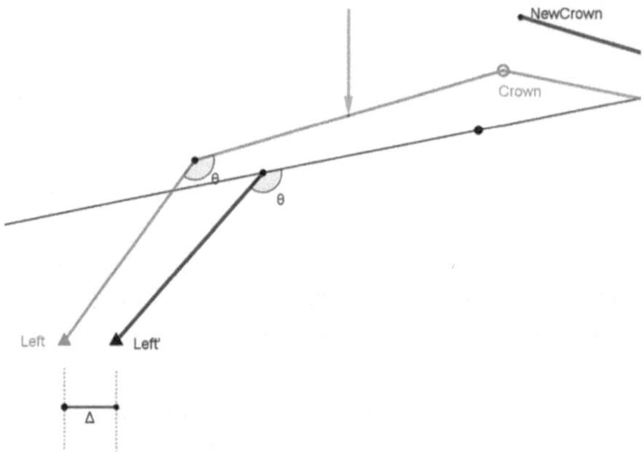

8.8
Moment carrying connections never change shape

Next, establish the second element of the arch in the perturbed configuration. This is quick to do via a circle centered on the perturbed bend point, whose radius is the length of the second arch segment. The intersection of this circle and the line which followed the pre-determined angle θ establishes the end of the second segment. This is shown in Figure 8.9.

The exact same process is repeated on the right side, and the two segments on the right side, with a controlled and fixed angle, will be created. What this will ultimately create is a Crown coming from the left hand two segment arch and a Crown coming from the right hand two-segment arch. The user will manually adjust

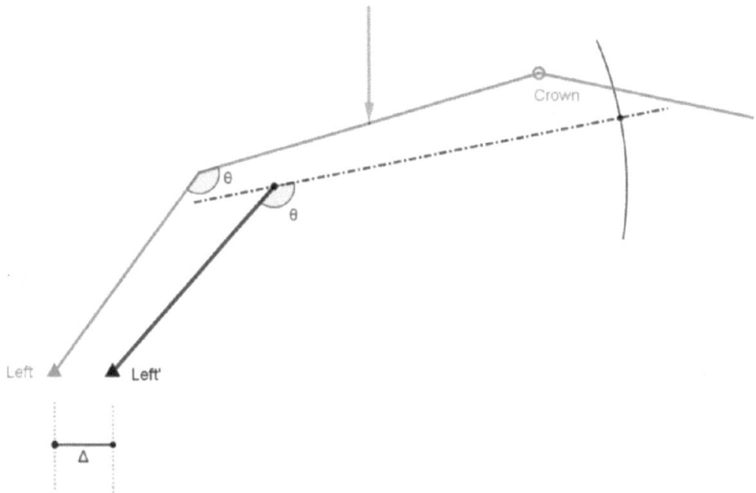

8.9
Establishing length of the second element, at proper angle θ

these two points until the error, which is the gap between the disparate Crowns, is very small. Ideally, these two Crowns should be at the exact same point, but a small error, or gap between the two, is perfectly acceptable. Alternatively, we could have built up the entire arch, Left to Right, and then we would manually adjust the Right support so that there is not much error between the Right' and the original Right. That second option will also be described in this chapter. For now, the reader should look at Figure 8.10 and see how the right side is built up.

In Figure 8.10, the fourth perturbed element is begun by creating a circle of radius Element4, centered on the unmoving Right support. The angle between the fourth and the third elements is established, and that angle is re-created on the perturbed element.

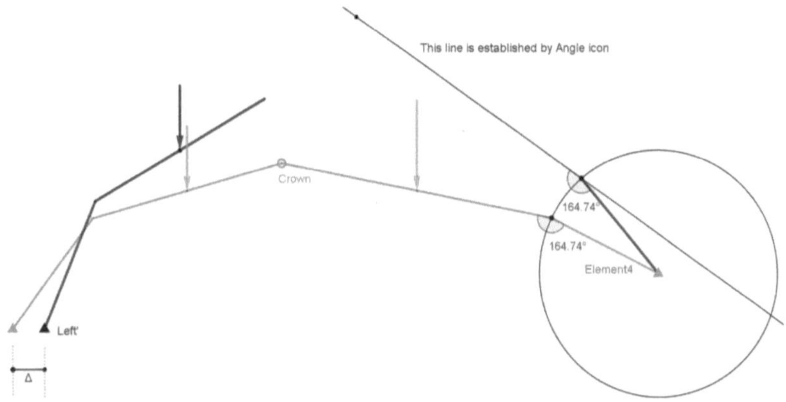

8.10
Establishing length of the fourth element, and second unchanging angle

Then, a circle of radius Element3 is centered on the right side moment carrying connection. The intersection of that circle with the line capturing the unchanging angle establishes the Crown from the right side of the arch. Clearly, the two Crowns should be at the same point, but the user will manually manipulate the geometry such that the distance between the two Crowns, which is the error in this construction, is very small. A large mismatch of Crowns is shown in Figure 8.11.

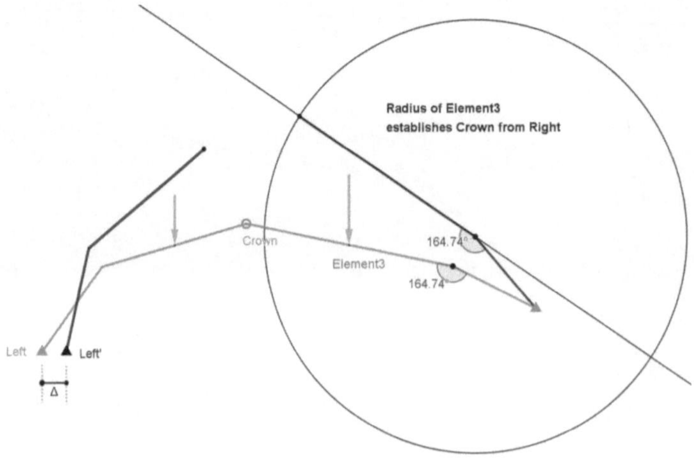

8.11
Establishing length of the third element, and the crown from the right side

Moving the bend points that denote the moment carrying connection moves the individual left and right two segment pieces of the arch, each as a rigid body. This allows the user to make the error, or the gap, between the two Crowns quite small. This is shown in Figure 8.12.

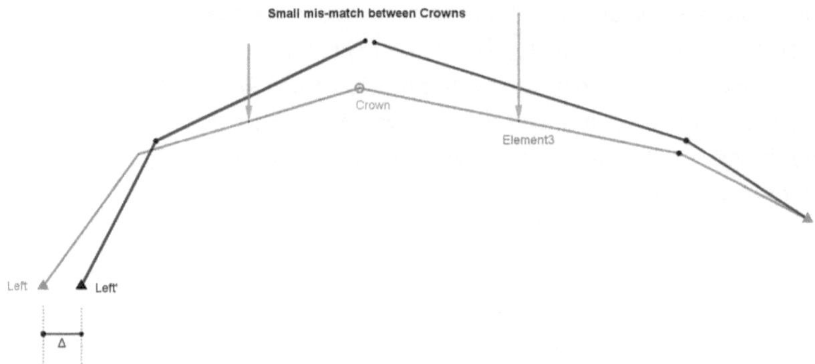

8.12
Closing the gap between the two crowns

Alternatively, the construction could have been completed sequentially, moving from Left to the Crown as before. Then, the third element, not the fourth element, would be found, with no restrictions on the angle between the second and third elements. Only the length is controlled, this is done by creating a circle, centered on the Crown which arose from the Left, a circle whose radius is Element3. These steps are shown in Figure 8.13.

Then, the angle at the unchanging right moment carrying connection must be established, which then is used to find the trajectory of the fourth element. This is shown in Figure 8.14.

Moving each of the rigid body two segment portions allows for manual adjustment until the error on the Right boundary condition is quite small. This is demonstrated in Figure 8.15.

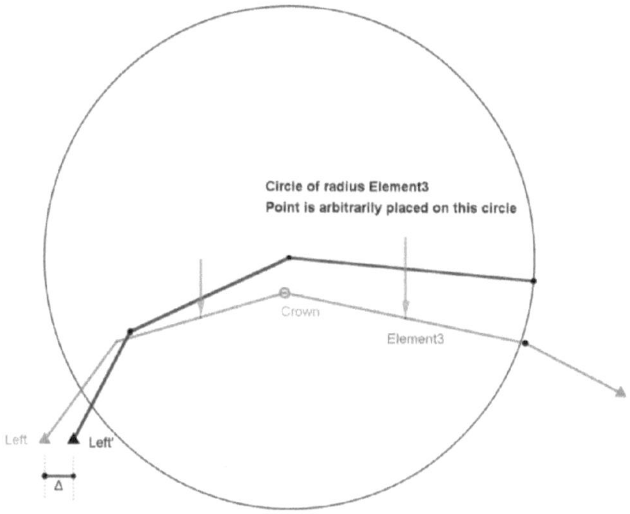

8.13
Recreating arch, solely from left to right

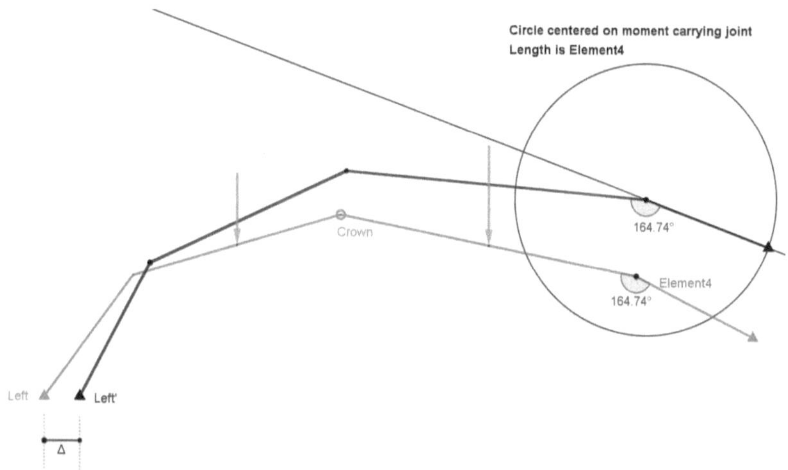

8.14
Fourth element established moving from left to right

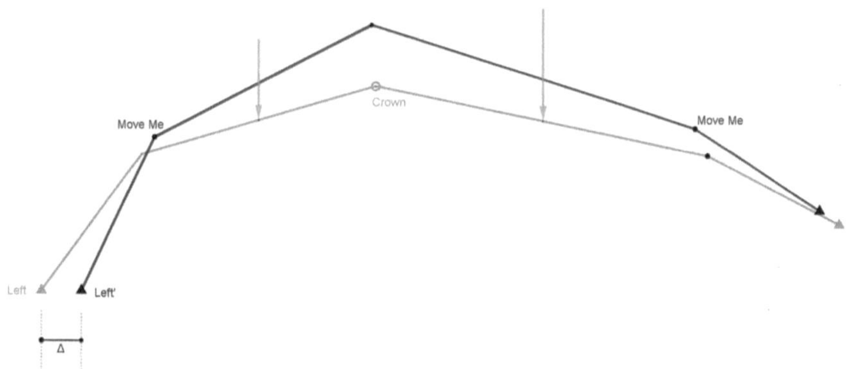

8.15
Making manual adjustments until error is small on right support

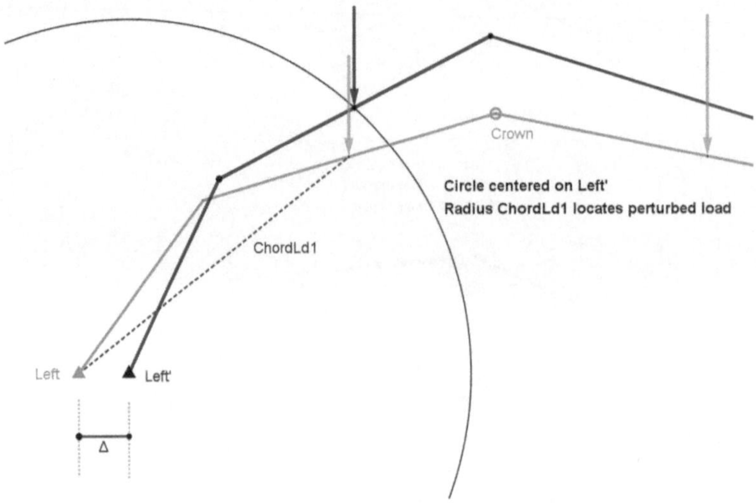

8.16
Chord length is key to establishing point location on a perturbed structure, a circle centered on Left'

Now that the construction is complete, the loads can be located in the perturbed configuration. As mentioned several times in the previous chapter, and has been demonstrated with previous constructions in this chapter, the chord length of the point, i.e. the shortest distance between the point of interest and the point about which it swings is all that is needed. Establishing that chord length is perhaps best done by a segment connecting the two points. Then, the name of the segment can be used for the construction of a circle, there is no need to use a command such as Length(segment), simply use the name of the segment.

A chord establishes the new location of the load on the right side of the arch. The circle established by this chord is centered on the Right' perturbed support. This is shown in Figure 8.17.

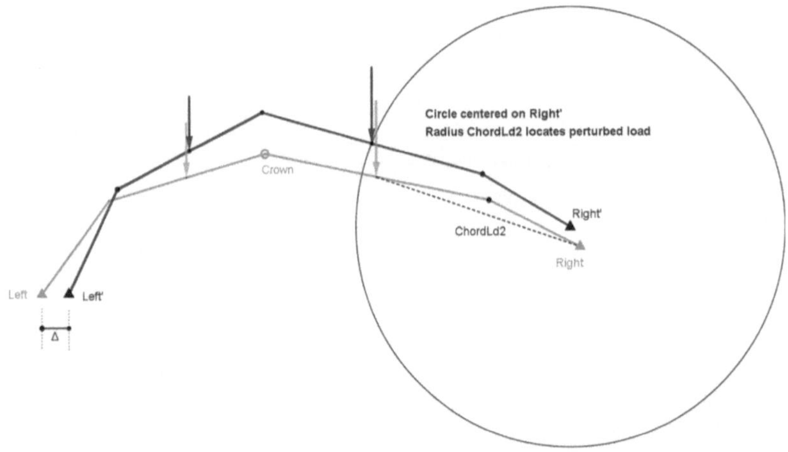

8.17
Different chord length is used to establish the second load point

We close this chapter with a revisiting of the first three-hinged arch shown in Figure 7.1 of Chapter 7, but now we seek the moment in the midspan of the left arch member. Details of the problem are shown in Figure 8.18. In Figure 8.18, the ability for the member to carry moment at the cut of interest is removed, and the

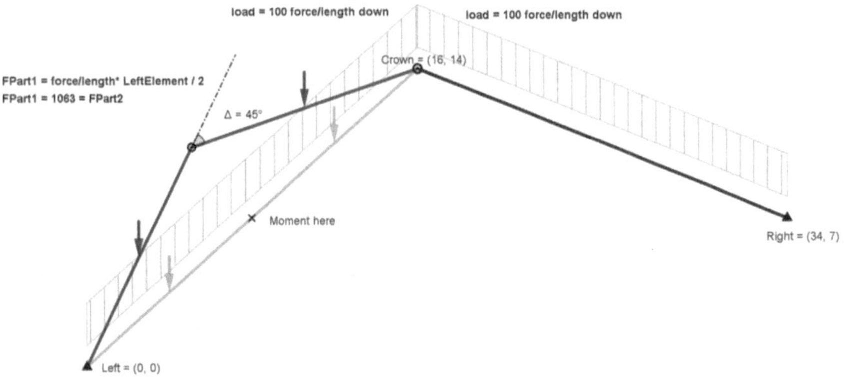

8.18
Large Δ to begin the study of the moment at midspan of the left element

perturbation Δ is the relative rotation of the single stretched member, on either side of the cut. Δ is purposely set to 45° in case the reader wanted to sketch a solution by hand, even though we are using the Modern Müller-Breslau Method, not the Classic Method.

As there is a distributed load that passes on either side of the kink formed by the cut, the load must be broken up as always, one part on one side, the second part on the other side. Figure 8.19 establishes the lofts and comes up with an answer for the moment, due to the large perturbation Δ.

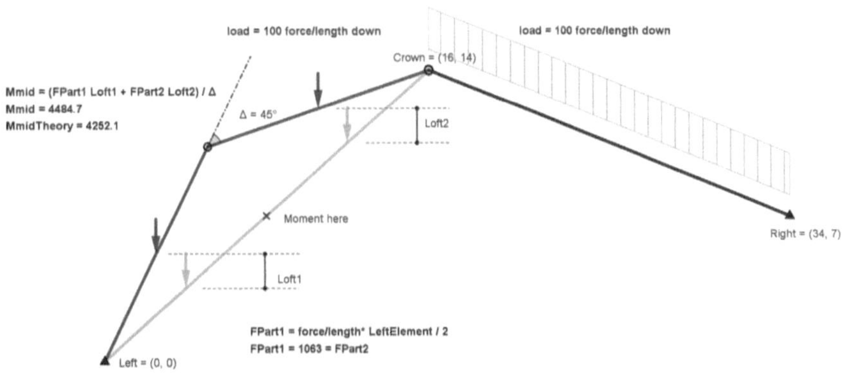

8.19
Lofts associated with large Δ to find the moment at midspan of the left element

Finally, as Δ gets small, the answer approaches the theoretical value, as shown in Figure 8.20.

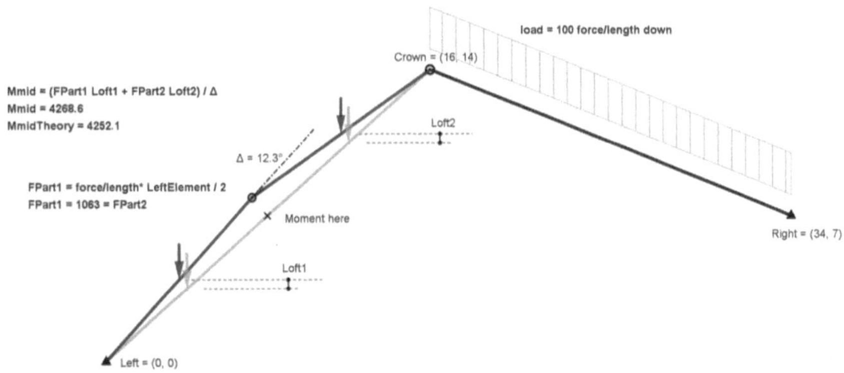

8.20
Lofts associated with small Δ to find the moment at midspan of the left element

GEOGEBRA FOR ARCHES ÉTUDES

The following studies or études should be done in GeoGebra.

Étude 8.1) A three-hinged arch is shown in Figure 8.21. It will be used for several études. It is a circular arc which is not an efficient structural form, but it is a good starting point for drawing. First, use the powerful argument of symmetry to deduce what the vertical equilibrating reactions at the Left and Right pinned supports must be. Second, simply estimate the magnitude and the direction of the horizontal equilibrating reactions at the Left and Right pinned supports are. These are known as thrust. Are they zero? Are they inward? Are they outward?

Third draw the structure in GeoGebra.

Hints: when drawing anything parametrically, resist the temptation to "hard-wire" anything that can change in a subsequent, but similar structure. For example, in this arch, it is perfectly acceptable to "hardwire" Left = (0,0). But don't do that for Right. Make Right a function of the known radius of the circle, Right = (Rad, 0) where Rad = 5.

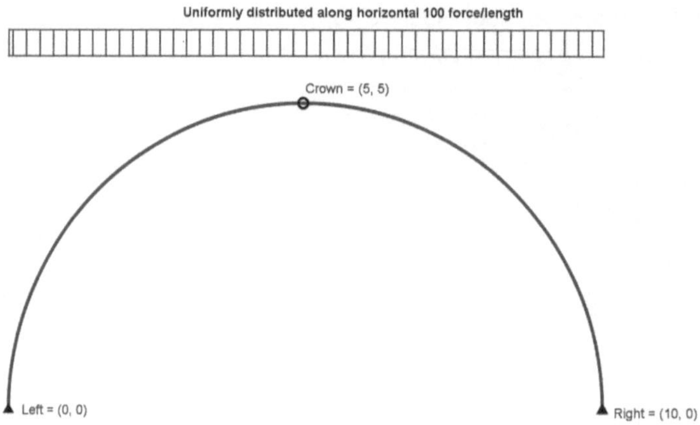

8.21
Circular arch setup

Étude 8.2) Use the Modern Müller-Breslau Method to quantitatively find the vertical equilibrating reaction at the Left pinned support of the structure shown in Figure 8.21. Make a very coarse model and simply split up the load into two pieces, half of the load goes to one side of the arch, half the load goes to the other side. Draw the perturbed arch for a $\Delta = 1$ used to find the vertical equilibrating reaction of the Left pinned support. Quantitatively find that vertical equilibrating Left reaction and compare it to the theoretical value that was predicted in étude 8.1. A suggested solution is shown in Figure 8.22. Other solutions may be more efficient and may make more sense to the reader. However, the reader can certainly use the hints provided in Figure 8.22. There, Left' = Left + $(0,\Delta)$. Then two chords are established, one from Left to Crown, one from Left to the point of the load. A circle of radius ChordCrwn centered on Left' is drawn and a circle of radius ChordCrwn centered on Right is drawn. Where those two intersect is Crown1, the lofted Crown for a perturbed Left vertical reaction.

A circle centered on Left' of radius ChordLd is drawn to find where the load on the left side is lofted to. A circle centered on Right of radius ChordLd is drawn to find where the load on the ride side is lofted to. The lofts are simply established as the difference of y coordinates of the load points. Figure 8.22 shows the solution for a large Δ = 1.

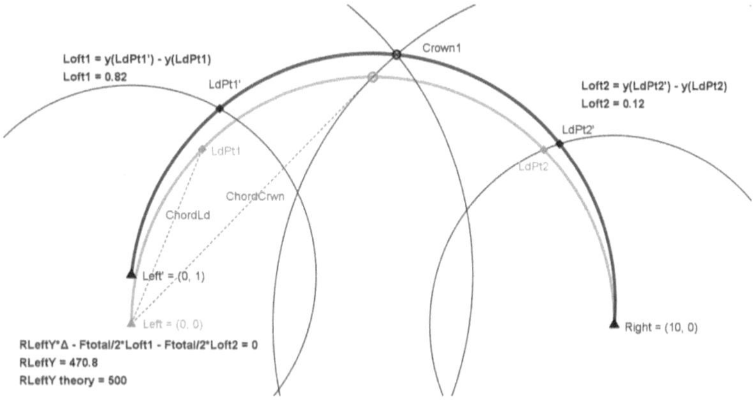

8.22
Solving for left vertical equilibrating reaction, large Δ

Figure 8.23 shows the same structure, solving for RLeft vertical for a small Δ = 0.2. Here the answer approaches the theoretical value of 500, i.e. one half of the total load, which makes intuitive sense.

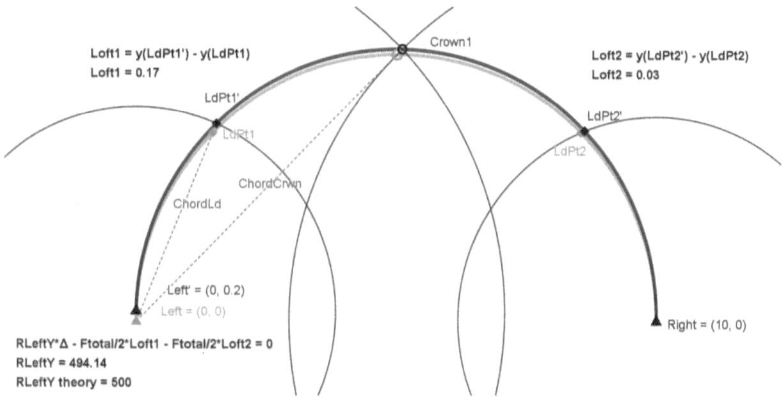

8.23
Solving for left vertical equilibrating reaction, small Δ

Étude 8.3) Use the Modern Müller-Breslau Method to quantitatively find the horizontal equilibrating reaction at the Left pinned support of the structure shown in Figure 8.21. A coarse solution for a large Δ = 1 is shown in Figure 8.24.

Étude 8.4) Use the Modern Müller-Breslau Method to quantitatively find the internal bending moment in the left "elbow", i.e. the point of bend on the left side of the arch in Figure 8.25. The Cartesian Coordinates of this "elbow" are (2,5). A coarse solution for a large Δ is shown in Figure 8.26. A more accurate solution for a small Δ is shown in Figure 8.27.

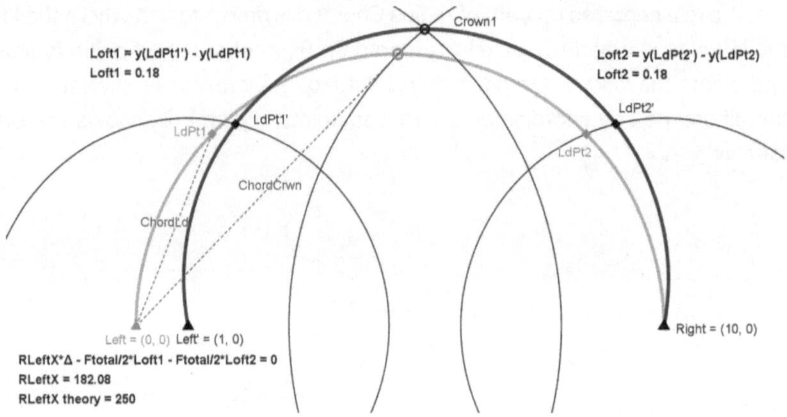

8.24
Solving for left horizontal equilibrating reaction, large Δ

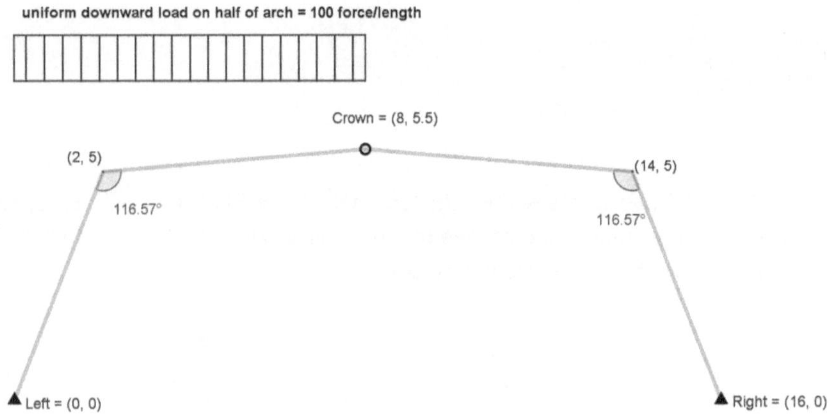

8.25
Setup for study, seeking internal moment at left "elbow"

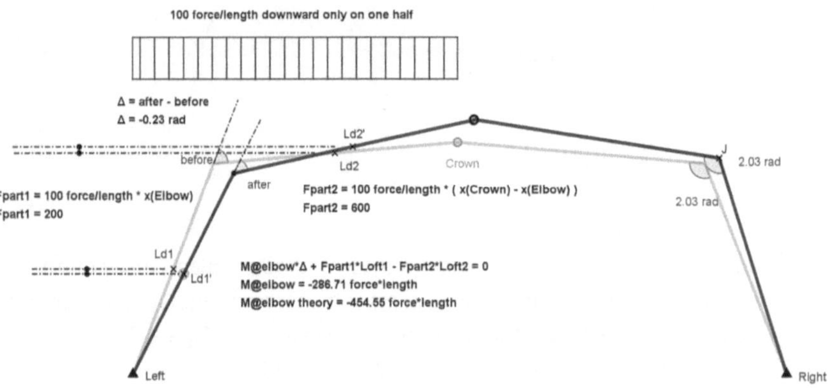

8.26
Solving for the moment at left "elbow", large Δ

8.27
Solving for the moment at left "elbow", small Δ

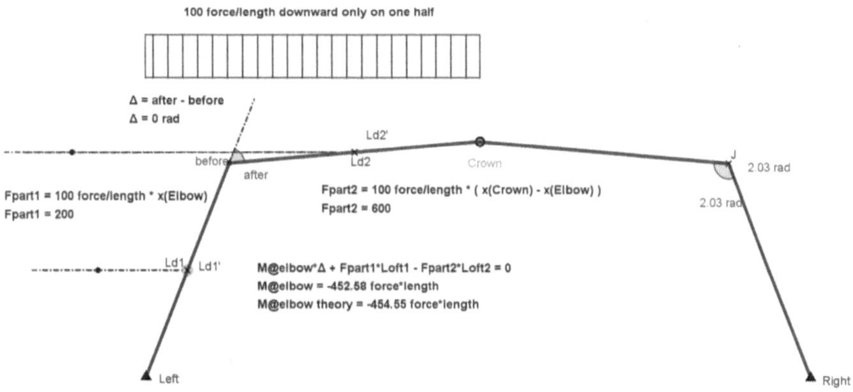

Étude 8.5) Use the Modern Müller-Breslau Method to quantitatively find the internal bending moment in the left "elbow", i.e. the point of bend on the left side of the arch in Figure 8.28. The Cartesian Coordinates of this "elbow" are (1.46, 3.54). A solution is shown in Figure 8.29.

8.28
Setup for study, seeking internal moment at left "elbow"

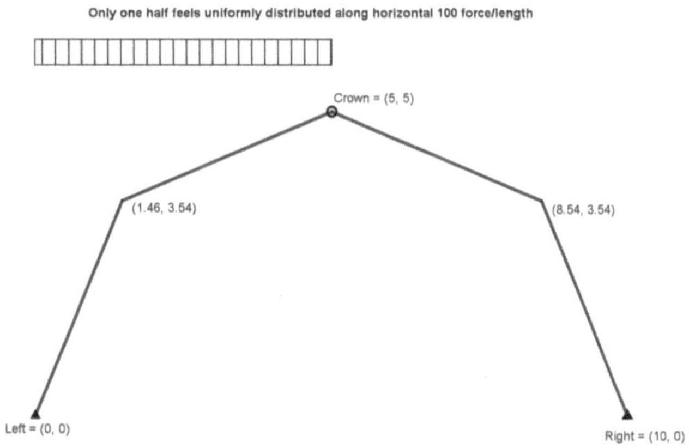

8.29
Solving for the moment at left "elbow"

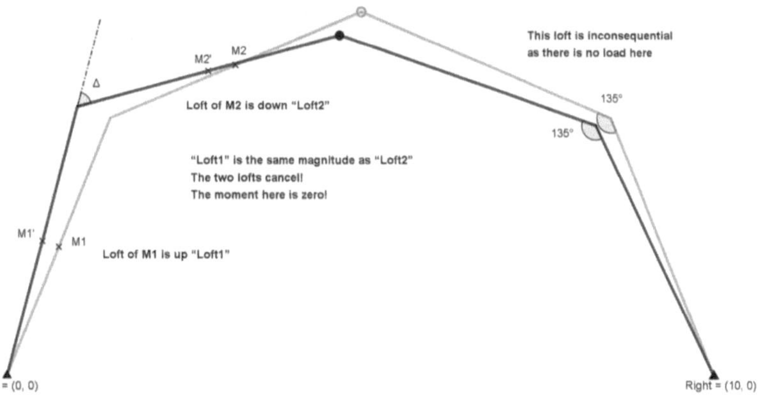

9 Indeterminate Structures

The most common indeterminate elements are beams that have moment carrying reactions, as well as force reactions at their ends. To achieve moment carrying capability in a steel connection, one needs to ensure that the top and the bottom flanges of the beam are linked to the subsequent element, such as a column. This allows the beam to transfer moment to the next element as shown in Figure 9.1 by pulling and pushing along the flanges, creating a moment or a "couple" which is defined as two forces equal in magnitude, opposite in direction, set some distance apart. In timber construction as well as in steel construction, a pinned connection is very common (Figure 9.2). Notice in Figure 9.2 that the beam is connected to

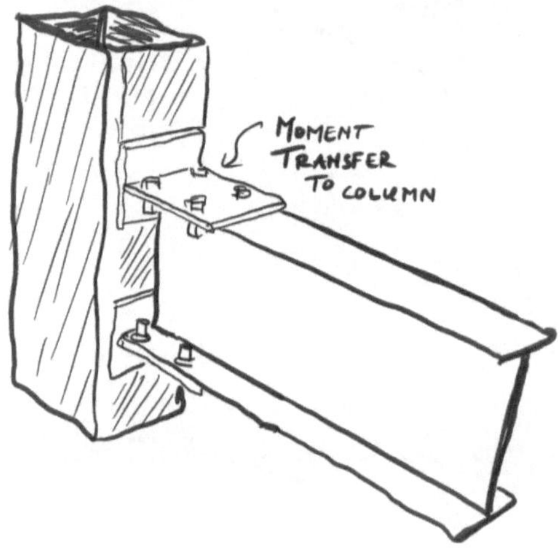

9.1
Moment carrying steel connection

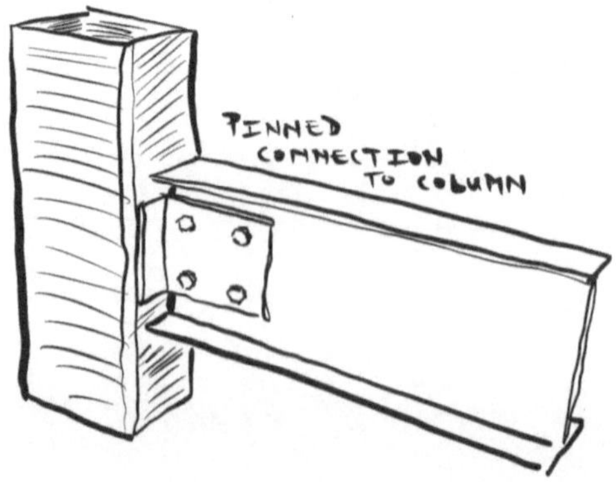

9.2
Non-moment carrying or pinned steel connection

DOI: 10.4324/9781003180913-9

the column only at the vertical central portion of the beam, known as the "web". Nothing on the top or bottom flanges engages the column, thus we assume that a zero moment is transferred from the beam to the column. This is a modeling assumption that is mostly true and is conservative, i.e. we ignore the small moment resistance that is actually provided and we assume that it is perfectly hinged to the subsequent member.

Reinforced concrete typically is considered "monolithic" meaning that elements flow into each other continuously, without break. This is because the reinforcing bars are placed such that they tie together adjoining elements, for example a column to a beam, then the liquid concrete is poured and vibrated so that it has no air pockets and it forms a seamless continuum. Clearly, such a connection carries moment and would cause an element to be indeterminate unless it was a cantilever. Figure 9.3 shows a flat slab connecting monolithically into a supporting wall. Reinforcing bars protrude, ready to engage the subsequent portion which can be "cast in place" or poured at the site.

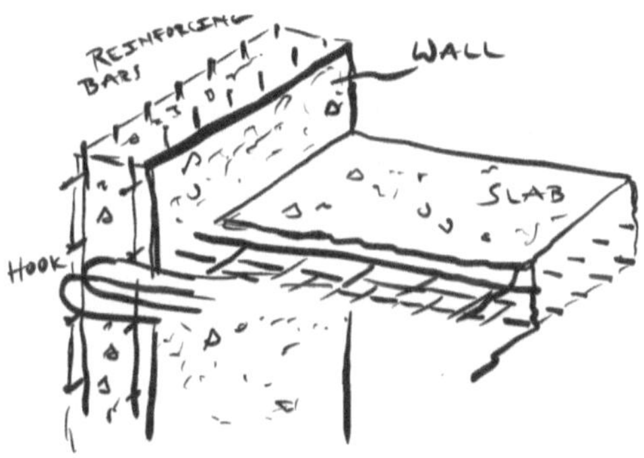

9.3
Moment carrying reinforced concrete connection

Structural designers use iconic symbols to designate moment carrying connections, versus non-moment carrying connections. These symbols have already been used throughout this textbook, a triangle for a pinned support that has two force reactions, one vertical and one horizontal. A roller for a support that only has a typically vertical reaction. And a wall that has two force reactions and one moment reaction. Indeterminate structures are much stiffer than determinate ones if all else is the same since they restrain rotation more. In a grid of beams, as was shown previously, a gap between members in plan view signifies symbolically a hinged or pinned connection.

In practice, people use sophisticated computer software to establish load flow in indeterminate structures. This is a reality of today's office environment. Yet, it is extremely important to manually check answers obtained by a computer, which we label as "theoretically correct values" in this textbook. Engineering students spend extraordinary time learning many classical methods to check such answers. But the Müller-Breslau (MB) Method brushes away all algebraic methods and dives deeply into a purely geometric construction that solely relies on Equation 1.1 of Chapter 1 yet again! The exact same equation is used in indeterminate structural analysis, but the premise is slightly different than before. Thus, the procedure for indeterminate analysis is as follows. We will be using the Classical Müller-Breslau Method for the study of indeterminate beams. This means that $\Delta = 1$ and Δ will not be reduced in the beam problems which are subject only to downward, gravity loads. It is hoped that future researchers will extend this method to more complicated loading scenarios. For this textbook we use the following steps for indeterminate analysis:

- Remove the capabilities of the structure to carry the Unknown of interest, be it an equilibrating external reaction or an equilibrating internal force or moment.
- Perturb the structure at the Unknown an amount $\Delta = 1$. This is a 45° relative rotation of elements for moment investigation.
- Watch as the applied loads are lofted due to the perturbation.
- Note that in indeterminate structures, almost every single element will be curved, rather than straight as was required in determinate analysis!
- Apply Equation 1.1 of Chapter 1 to find the Unknown.
- Unless the perturbed shape is exactly correct, the answer will be approximately, not exactly, correct.

There are advantages and disadvantages of the MB Method when used on indeterminate structures. The overwhelming advantage is that the procedure can be sketched on a table napkin and reasonable answers can be found using the sketch. A huge advantage is that the method can be learned quickly, without any algebra or any sophisticated theories. Another major advantage is this:

- The perturbed shape has absolutely nothing to do with the applied loads!

This point needs to be emphasized and elaborated upon. The perturbed shape is due to only the Unknown moving through Δ. For example, in a beam that is fixed at its left end and fixed at its right end, the perturbed shape induced by removing the Left Wall Moment reaction and applying a Unit Rotation (i.e. 45°) is shown in Figure 9.4. It has nothing to do with any applied loads, none of which are shown in Figure 9.4.

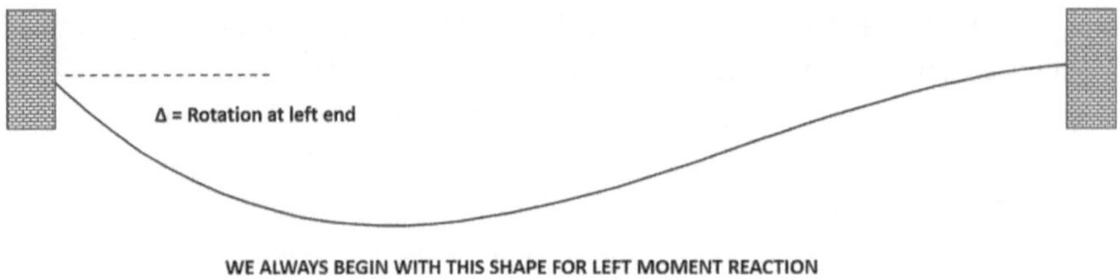

9.4
Fixed indeterminate beam seeking the left moment reaction

It would be unreasonable to expect a novice reader to understand how the beam of Figure 9.5 would deform due to any possible applied load. This is never ever needed in the MB Method.

However, we contend that it is perfectly reasonable to walk even a novice reader through the perturbed shape of Figure 9.4 and to do so with quick, easy to remember rules that can be sketched by hand. The experienced reader may be very skeptical about this position from a pedagogical point of view. Yet, we urge such readers to

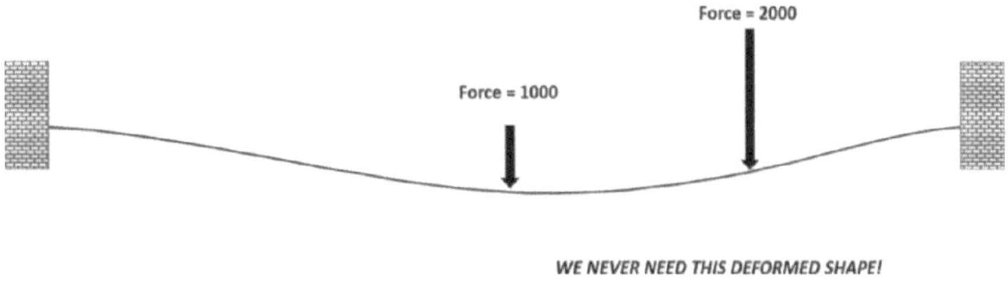

9.5
Fixed indeterminate beam deformation due to loads, never needed

go back to their student days when they first saw a deformed shape of a beam. Somebody had to guide all students seeing such forms for the first time. Similarly, the experienced reader may well recall that in matrix methods courses or courses that used the Slope-Deflection Method, the displaced shape due to a single perturbation (then called a degree of freedom) was something that was eventually learned and that became very useful. This leads to a re-statement of the major advantage:

- The perturbed shape is due to the Unknown moving through Δ, not due to any load at all. Furthermore, we consider this Δ to be a unit rotation when seeking moments.

Our pedagogical position is that it is extremely useful for all readers to thoroughly recognize a few classic and common perturbed shapes. We will focus only on a few such classes of structures in this chapter, but readers are strongly encouraged to find other common typologies and to apply the MB Method to these new cases. In fact, this chapter is the most theoretical one of the book, and a few newly minted ideas will be presented to encourage readers to advance the method on their own. The *Equilibrium Without Statics Revolution* has begun and you are part of it!

Restating the fundamental premise of this chapter, the perturbed shape for a particular problem can be presented *a priori* to the analyst. The analyst learns this shape and applies it for that class of problem, for any applied load. And the perturbation Δ is unity, i.e. one when seeking a moment.

This leads to the major disadvantage of this method. One must use a ***prescribed shape*** that is not easy to intuit by a novice. The novice reader must trust that this shape is correct. Furthermore, for all readers, the shape must be easy to remember. Note again, if this shape is not exact, then the answer will be incorrect. If, perchance, the shape is exactly correct, then the answer will be exact. But in this textbook, it was decided to embrace a fairly easy to remember approach to defining a perturbed shape, rather than exactitude. As such we focused only on only a few classes of problems:

- Propped cantilever, i.e. fixed at one end and a roller support at the other end.
- Continuous, two span beam.
- A single bay, rectangular frame with pinned supports at the bottom of each column.

Of course, an infinite variety of other cases exist and there is where the readers are encouraged to do creative research using this method.

With each of these cases, we will look only at a support movement, or a unit rotation. A support movement will provide an answer for an Unknown equilibrating force reaction. A unit rotation will provide an answer for an Unknown equilibrating moment. Once enough of these are solved, the problem becomes determinate and the previous techniques can be used to continue studying any problem.

It must be emphasized to the reader that this is an approximate method that is meant to teach load flow. From a research point of view, it is interesting to see if we can sketch out perturbed shapes accurately for any type of indeterminate problem. Even far-fetched scenarios seem within grasp of an approximate answer, provided we have a visual database of perturbed conditions. For example, a somewhat fanciful scenario is an indeterminate plate, subjected to multiple in-plane loads, we seek the vertical reaction at the lower right corner, labeled as Reaction 1 in Figure 9.6. Notice the wide variety of downward loads applied in the plane of the plate, along its top surface.

If this seems impossible to solve geometrically, recall that the perturbed beam subjected to a single-end rotation may have seemed impossible to predict at one point in the reader's training. Figure 9.7 shows a solution for the perturbed shape of the lower right vertical reaction, Reaction 1, being perturbed a unit amount.

Studies such as the one shown in Figures 9.6 and 9.7 will not be pursued here, but are left as enticing trails to follow for the curious reader.

Yet several complexities must immediately be grappled with, beyond the pesky unit rotation of moments. One is the fact that even for the simplest indeterminate structures, stiffness variations are inevitable and must be accounted for. For example, in a rectangular single-bay single-story frame, the columns and the beam

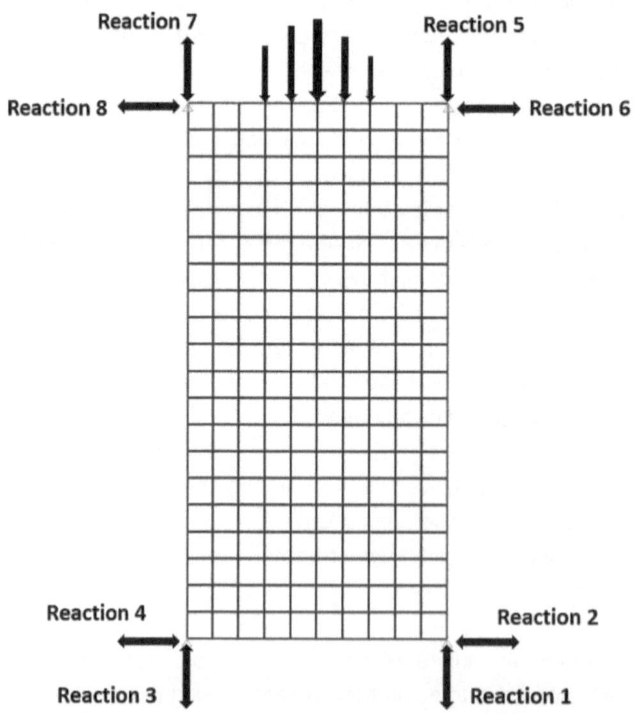

9.6 Indeterminate plate, loaded in its own plane, seeking Reaction 1

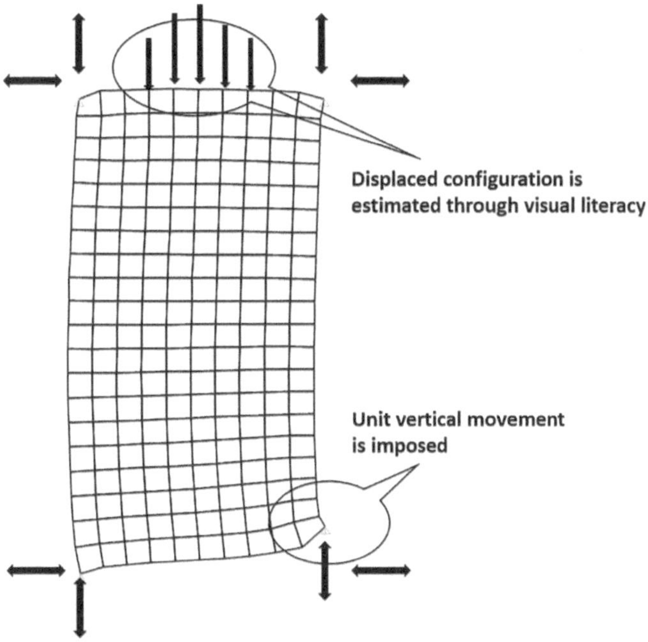

9.7
Geometric solution for Reaction 1

have varying stiffness since often the size of the frame is not perfectly square, or because the elements themselves vary, i.e. if the cross section of the beam is different than the cross section of the column. The difference between a rectangular frame with a very stiff beam and very flexible columns, versus the same frame with a very flexible beam and stiff columns is shown in Figure 9.8. In Figure 9.8, notice that we are seeking the left horizontal reaction. The perturbed shape has absolutely nothing to do with any applied loads, and no loads are shown in Figure 9.8. It is this premise that is our fundamental precept in this chapter, learn a specific perturbation and apply it to a particular class of problem, here a one bay rectangular frame with pinned supports. That is what is meant by "visual literacy", as it applies to load flow in building structures. This literacy can indeed be cultivated by guided explorations such as those highlighted in this textbook.

Figure 9.9 shows another class of problem we will study in this chapter, namely the propped cantilever. The fundamental premise of the MB Method is that the shape is known *a priori*, and that it is relatively easy to sketch by hand. There are

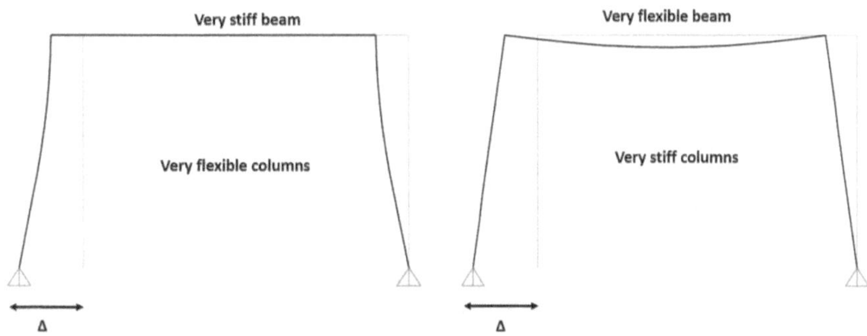

9.8
Pinned base indeterminate frames, seeking the left horizontal force reaction

five key points in the MB Method for the propped cantilever. One is the left end of the beam. Then three MB points, MB1 MB2 and MB3 are established using a 45° rule which will be explained shortly. The fifth point is the right end of the beam. In Figure 9.9, the three MB points are shown with curves fitted through them, as well as through the Left and Right points. Solid circles represent theoretical values of where the perturbed beam actually might be for a typical cross section, span and material property such as found in a building. Exactitude is not the goal, we seek a simple curve that is easy to remember and can be sketched by hand.

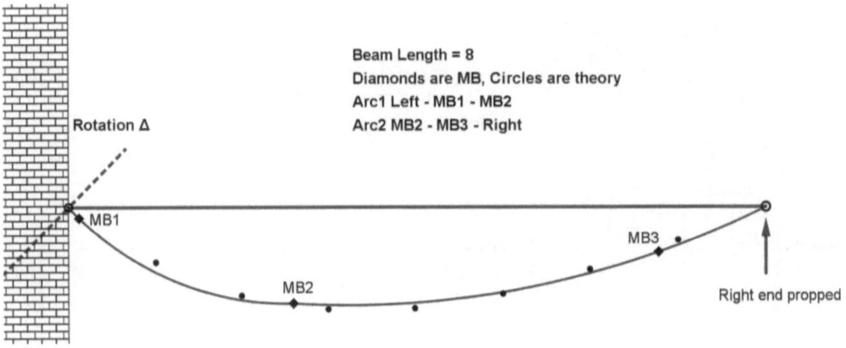

9.9
Propped cantilever indeterminate beam, seeking the left moment reaction

Students often are confused by the apparent paradox of "small angles" and "unit angles". The construction used for the propped cantilever of Figure 9.9 starts off with a so-called "unit" rotation of 45°. Note that 45° is not 1 radian, so the reader is urged to focus on this slight paradox. A unit rotation is the essence of the Classical Müller-Breslau Method, and Müller-Breslau himself used a 45° rotation so that the "rise" is equal to the "run" in his so-called "Influence Line" geometric constructions. To help resolve this slight paradox, one can think of this perturbation as a zoomed in close up of a very small, or nearly "virtual" rotation. Or one can simply think of it as a unit rotation.

Here, we assume that this is a unit rotation and when we divide by Δ in Equation 1.1 of Chapter 1, we have divided by unity. Thus we will use Equation 9.2 for the propped cantilever studies.

$$Unknown \cdot \Delta + \sum Force_i \cdot loft_i = 0 \quad \text{(Equation 1.1) of Chapter 1}$$

$Unknown \cdot 1 + \sum Force_i \cdot loft_i = 0$ (Equation 9.2) for beams subject to gravity only loads.

If the circles, which arose from theory, would have fallen precisely on the perturbed shape sketched in Figure 9.9, the answer for the Unknown Left moment would be exact. We argue that this approximate curve will help instill in the reader an understanding of how load flows in indeterminate structures. We seek to show how particular external loads flow to support reactions, or in some cases how they have little influence on the Unknown being sought. The construction of the perturbed shape due to a rotation of the fixed wall support is very simple indeed. A 45° line is drawn at the Unknown point of moment reaction in the fixed wall. That angle is bisected to form ½ of 45˚ or 22.5°. The same 22.5° angle is drawn on the roller side of the beam. One MB point lands on the 22.5° axis close to the roller, another MB point lands on the 45° axis close to the wall. The intermediate MB point lands at the intersection of the two 22.5° lines. Circular arcs are passed through three successive points. For example, one circular arc is passed from Left, through MB1 to MB2. That arc appears very straight in Figure 9.10, nevertheless, it is a circular arc, not a straight line. Another circular arc is passed from MB2 through MB3 to Right. Figure 9.10 shows this construction with the wall on the right end of the beam. A discontinuous slope is an unavoidable product of this simplified method, i.e. there is no real need for slope continuity between the two arcs in this simplification.

9.10
Rules for the perturbed shape of a propped cantilever

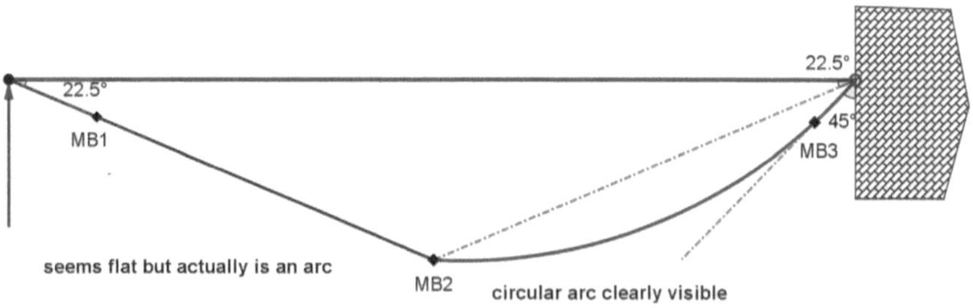

Figure 9.11 shows the construction via a simple hand sketch, this time with the wall on the left end of the beam. A hand sketch would simply spline the five points.

9.11
Perturbed shape of a propped cantilever via hand sketch

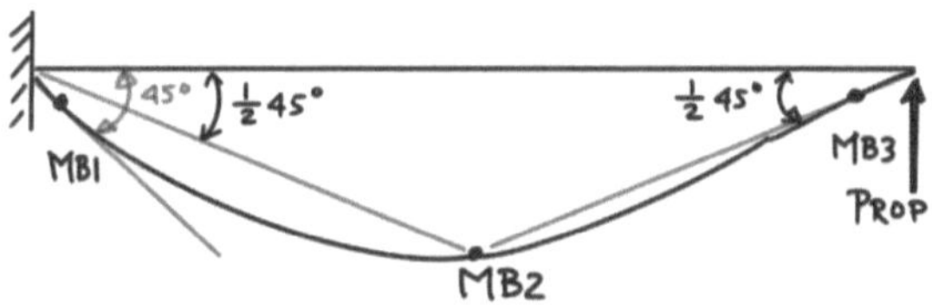

Is this construction valid for absolutely any beam that is fixed/pinned? No. But for building structures, this method appears to perform fairly well. As with each and every example in this textbook, the reader is encouraged to explore and to improve the methods presented!

For now, a few examples of indeterminate beams will demonstrate the power of the method.

Figure 9.12 shows a propped cantilever subjected to two point loads. The perturbed shape was constructed via the rules established by Figure 9.10. Agreement with theory is remarkable.

The same problem is solved by a simple hand sketch. That sketch is shown in Figure 9.13.

A slightly more complicated propped cantilever is shown in Figure 9.14. Here, the beam extends past the prop support and that portion of the beam is uniformly loaded. There is also a concentrated force between the wall and the prop support. We seek the moment in the Left Wall. The theoretical values are section dependent, thus a typical structural section was used here.

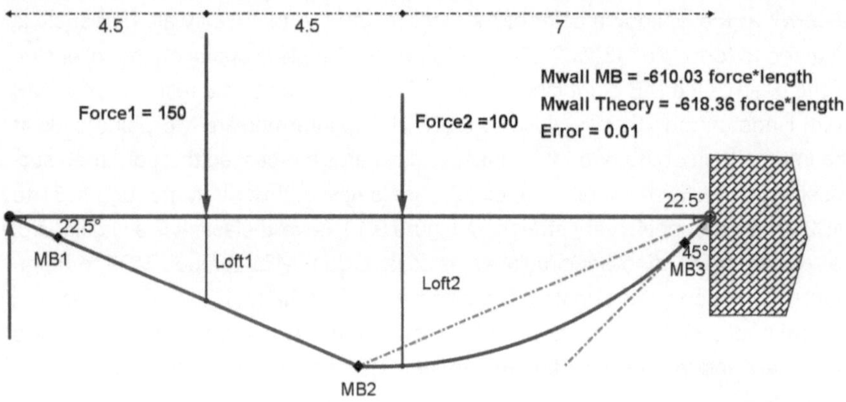

9.12
Solution for Mwall in a propped cantilever

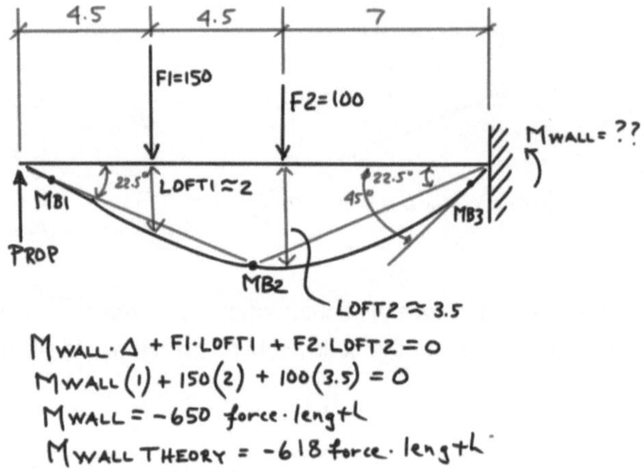

9.13
Solution for Mwall in a propped cantilever via hand sketch

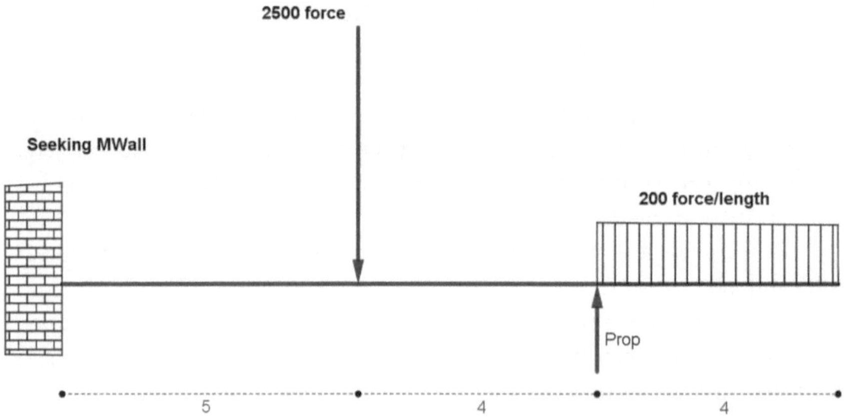

9.14
Seeking Mwall for a more complicated propped cantilever

The student may be tempted to draw the perturbed shape with a bent overhang, but this would be erroneous as the perturbed shape has nothing to do with the applied loads, it is induced solely by the rotation at the Unknown Left Wall equilibrating moment. The elementary rules of construction from Figure 9.10 are applied to the beam in Figure 9.14 and the beam remains straight to the right of the prop support. The lofted configuration is shown in Figure 9.15.

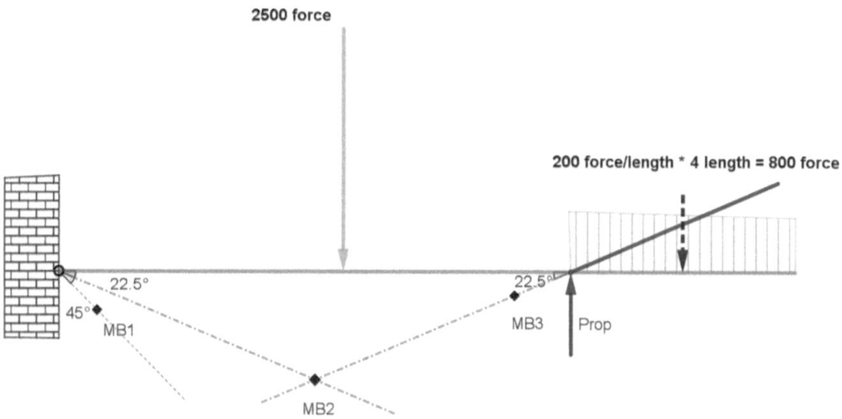

9.15
Seeking Mwall for a more complicated propped cantilever

To actually construct the perturbed shape, two circular arcs are drawn as was done previously. One arc passes from Left through MB1 to MB2. A second arc passes from MB2 to MB3 to the Prop. That curve appears linear in Figure 9.16, but it is in fact slightly curved.

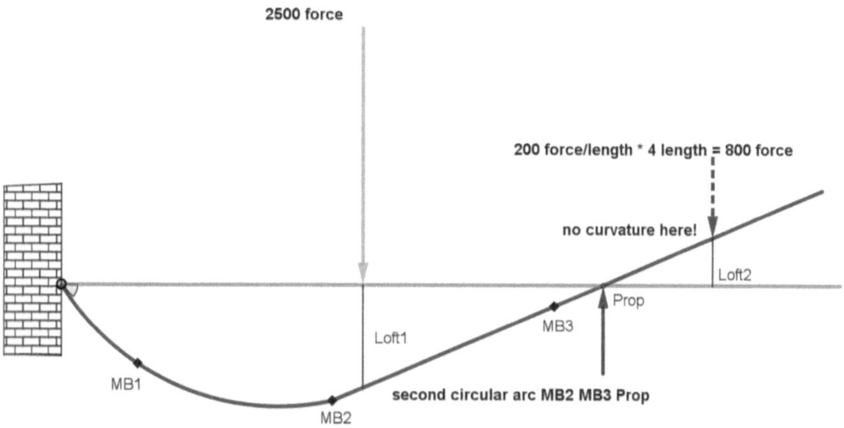

9.16
Establishing the three MB points and two circular arcs

The moment is found by Equation 1.1 of Chapter1 with $\Delta = 1$. As described earlier, this is the Classical Müller-Breslau Method, thus we divide by 1 not by the angle in radians. Notice that the 2,500 force does positive work and the equivalent concentrated force to the right of the prop does negative work. Having such a geometric tool allows one to instantly assess whether a particular force adds to, or detracts from, the equilibrating moment in the wall. Figure 9.17 shows the completed solution.

The error is quite large here, but still under 10%. Again, the theoretical value depends on the section chosen, here a typical section was used.

The next example shown in Figure 9.18 is an indeterminate beam with two spans. The MB Method can be used to find one of the reactions. A circular arc is passed through three points, Left, Prop and then Right. Here, the loft is very large, to clearly see the perturbed beam. A variety of lengths are explored, to show that the method is independent of Metric or Imperial Units, and that it works for a wide

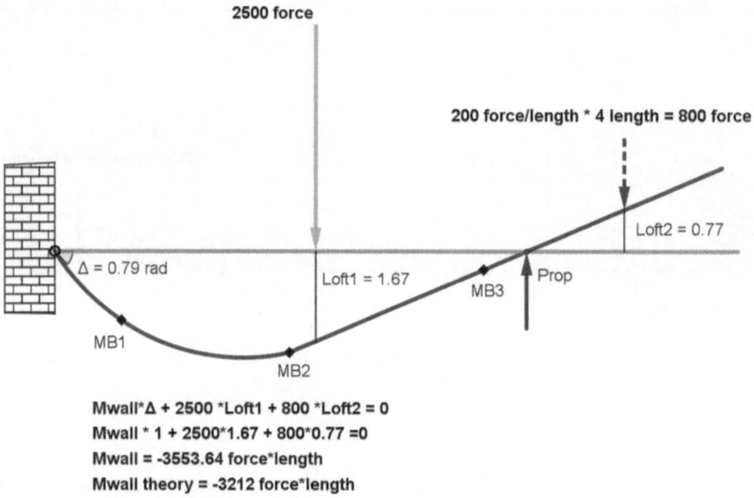

9.17
Approximately solving for Mwall

variety of problems. This unlocks the problem which becomes statically determinate. We can consider these shapes to be "funicular," not in the classical sense of a hanging chain, but more in the sense that they are the shapes that follow the imposed boundary conditions. If the shape is correct, i.e., if it is "funicular," then the MB Method gives exactly correct answers for equilibrating forces or moments. As stated in Chapter 1, this allows for only external work to be considered in Equation 1.1. If the shape is slightly incorrect, then slightly inexact answers will ensue. Three funiculars for a two-span indeterminate beam are shown, each one attempts to capture the left vertical reaction. The variations of stiffness of the first span versus the second span are investigated. Notice the beams of Figures 9.18–9.20 all are 36 units long, but the spans vary.

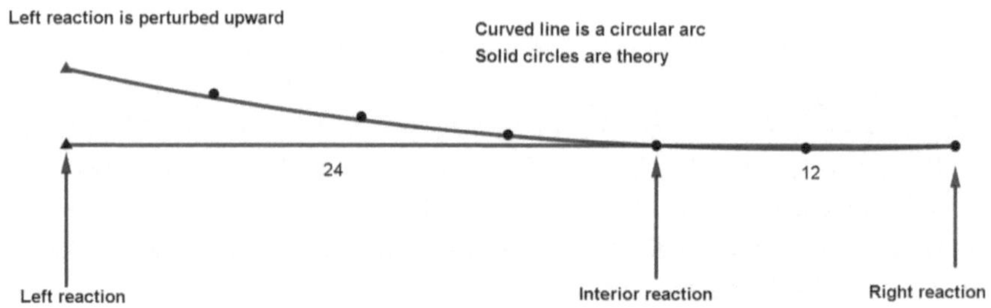

9.18
Left span is more flexible

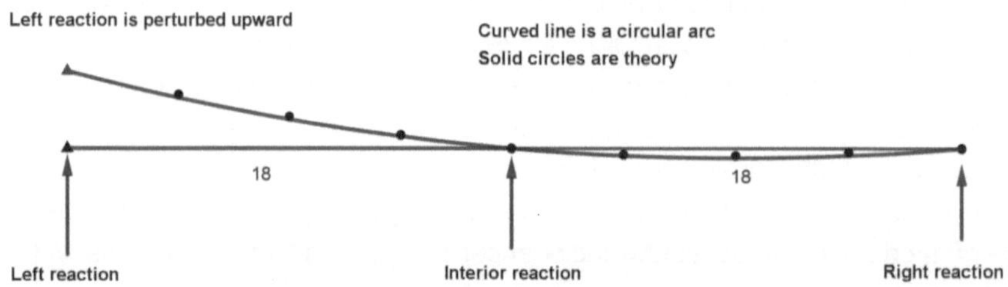

9.19
Both spans have the same stiffness

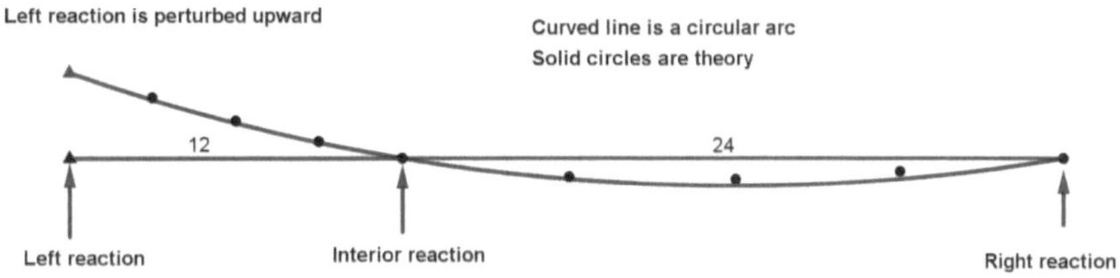

9.20
Right span is more flexible

An alternative approach is to simply find the moment in the beam above the interior prop support. This alleviates the analyst from finding external vertical reactions, which is the spirit of the previous indeterminate studies. Recall that these previous studies always found a moment, never a shear.

We begin by studying a symmetric problem, a prismatic beam with an interior prop and matching spans on either side of the prop. Prismatic means that the cross section does not change shape along the length of the beam. We seek the moment in the beam directly above the centered prop of the prismatic beam.

One way to think about the perturbation is to recall the previous propped cantilever, the fixed support rotated 45°. The top of Figure 9.21 shows the rotation on either side of the cut stylized as 45° rotations of fixed supports, which do not exist in our problem, the beam is continuous over the prop in our problem. The bottom of Figure 9.21 shows the correct angle, which is 45° relative across the cut being studied.

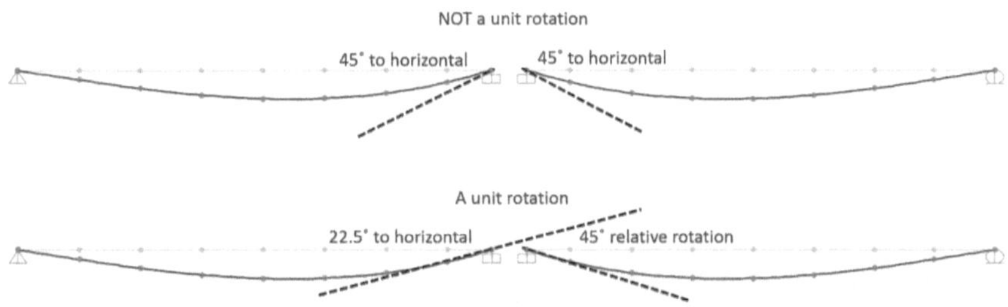

9.21
Studying a unit rotation at midspan of a propped two-span beam

Notice that in this situation shown in Figure 9.21, either span is essentially the same as the previous Fixed-Pin single-span case studied earlier. Previously, the difference was the slope of the single span versus the difference of the fixed support which was a 0° slope.

Now, in the two-span, symmetric case shown in Figure 9.22, we seek a net of 45°, thus each portion rotates 22.5° individually. This explains why the Δ in the Müller-Breslau Equation is still 0.785rad (45°) or Unity, it is this Δ that drives the entire perturbation. We use the same ideas as the fixed-pinned case, but we divide all the construction angles by 2. The three MB points are immediately located, even on a table napkin sketch. Of course, the spline is not perfect, but the simplicity of locating only three MB points may pedagogically override the weaknesses of an imperfect spline. As previously stated, if the spline were perfect, it could be considered "funicular" and thus would give exactly correct answers.

Now, consider a uniformly distributed load over only the first span of a two-span indeterminate beam. We seek the equilibrating moment in the beam directly

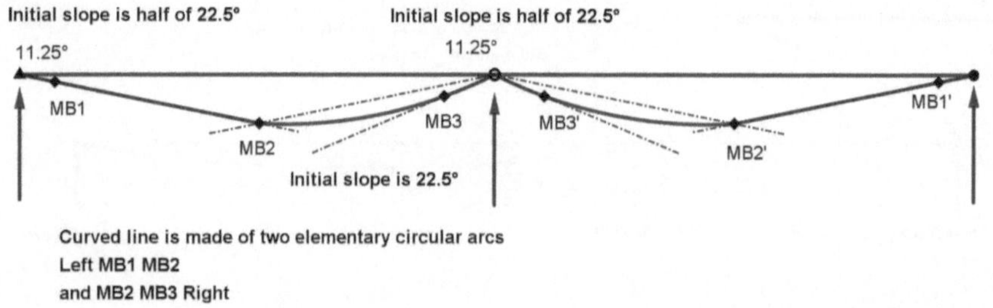

9.22
Perturbation to find M@prop in a two-span symmetric beam

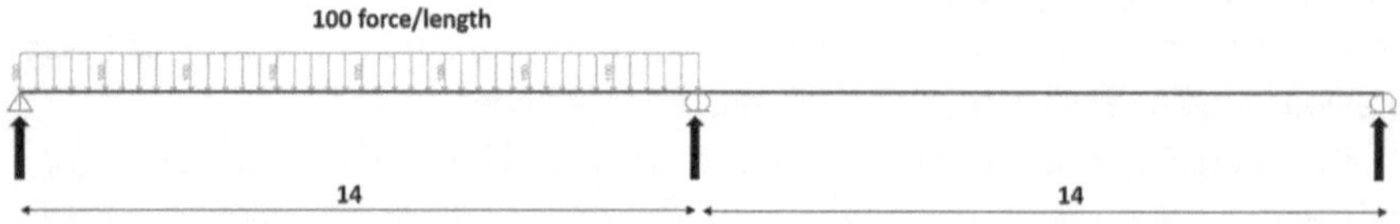

9.23
Seeking M@prop in a two-span symmetric beam loaded on one half

above the interior support, also known as a "prop". The problem setup is shown in Figure 9.23.

Recall once again that the deformed shape due to the applied loads is never ever needed in the Müller-Breslau, only the perturbed shape due to the Unknown being sought. To reinforce this point, the deformed shape for loads of Figure 9.23 is shown in Figure 9.24. Of course, they are wildly different than the needed perturbation of Figure 9.22.

9.24
We never need this deformation due to applied loads!

Using the Müller-Breslau construction to find the moment at mid-span for a symmetrically loaded beam is quick and easy. The error, even for a coarse distribution of the distributed load and an imperfect perturbed shape, is a remarkable 1%. Note that the load must be broken up in the left span, as there are "kinks" everywhere in this span. This is shown in Figure 9.25.

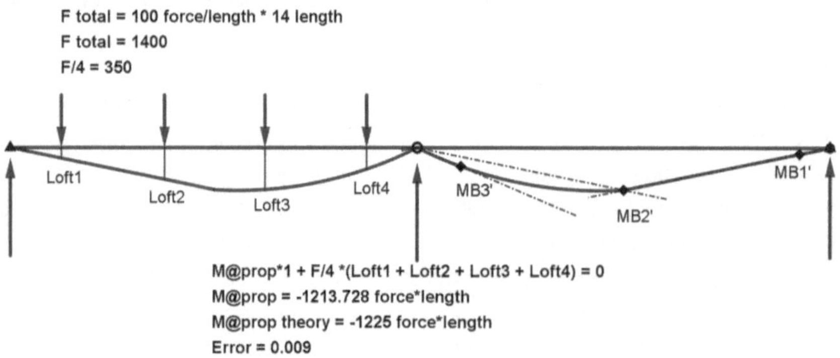

9.25
Solving for the moment above the interior prop

Since stiffness variations are an essential part of the indeterminate analysis, the following example suggests one way of accounting for stiffness variations. Other methods may be better, but the proposed method is easy to use and simple to remember. To set up the problem, consider a two-span indeterminate beam, subjected to a uniform load only on the longer span. This is shown in Figure 9.26.

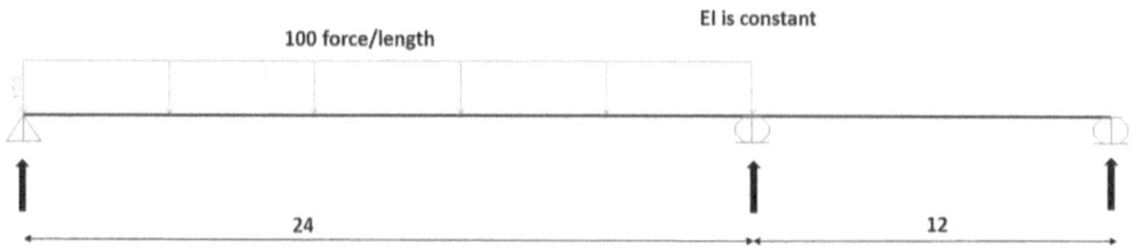

9.26
Asymmetric problem, hence varying stiffness

If we were to ignore the fact that the shorter span is stiffer than the longer span, we would begin with an incorrect perturbation for the moment in the beam above the interior prop as shown in Figure 9.27. This would be incorrect as it neglects the variation of stiffness of each span. The height formed by the net 45° angle, or the individual 22.5° angles is important and noteworthy.

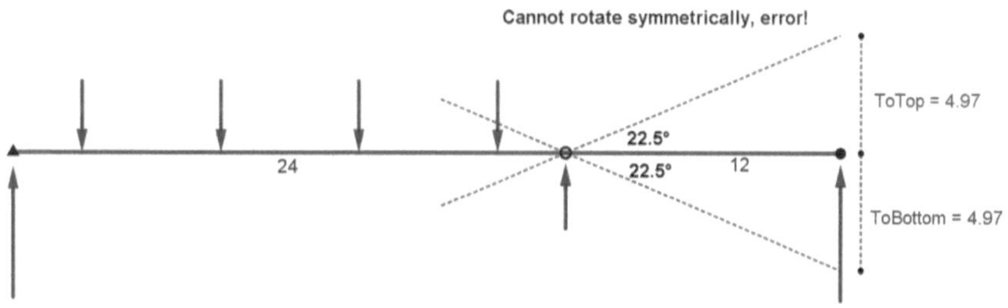

9.27
Incorrect perturbation that ignores stiffness variation

The construction of Figure 9.27 is incorrect because of the stiffness variations. Yet it is a useful starting point to explain the correction needed to adjust for the varying stiffnesses of the spans. A needed term for the adjustment due to stiffness variations is the Height formed by the 45° where it intersects the right propped support. Here, that Height is 2*4.97 = 9.94 for the 12 unit span. (The careful reader will notice that herein lies one of the paradoxes of the "unit rotation", namely that an arc "s" would be equal to radius "r" as in $s = r\theta$ if θ was 45°, however here the Height of the vertical straight line is 9.94 for the r of 12.)

To account for the stiffness variations, we use the ratio of stiffness on either side of the MB cut. The term "bottom" will refer to the distance from the support to the 6 o'clock mark on the previous "Height" segment. The rule to use is:

$$bottom + stiffness\ ratio \cdot bottom = Height \quad \text{(Equation 9.3)}$$

$$bottom \cdot (1 + stiffness\ ratio) = Height \quad \text{(Equation 9.4)}$$

$$bottom = \frac{Height}{1 + stiffness\ ratio} \quad \text{(Equation 9.5)}$$

$$stiffness\ ratio \geq 1 \qquad \text{(Equation 9.6)}$$

Recall that "bottom" is the distance from the support to the "bottom" i.e. to what was 6 o'clock on the previous Height segment. The distance to the top is then:

$$top = Height - bottom \qquad \text{(Equation 9.7)}$$

So what is the "stiffness ratio"? As is the case for many issues in structural engineering, this term can be the subject of academic debate. We choose to define stiffness as:

$$\frac{E \cdot I}{Length\ of\ Span} \qquad \text{(Equation 9.8)}$$

Where E is the modulus of elasticity, units are force/length^2, I is the moment of inertia, units are length^4, and Length of Span is the length of the span on one side of the cut, obviously, units are length. This definition does not have the traditional units of stiffness which are force/length, but since we are dealing with rotations here, and since we are aiming for ease and simplicity, we use this basic definition. Then for the problem of Figure 9.27, since EI is constant, we have the stiffness ratio as:

$$Stiffness\ Ratio = \frac{EI/12}{EI/24} = 2 \qquad \text{(Equation 9.9)}$$

Note that we required the stiffness ratio to be greater than, or equal to 1, so we placed the EI/12 in the numerator.

$$bottom = \frac{Height}{1 + stiffness\ ratio} = \frac{2 \cdot 4.97}{1+2} = 3.31 \qquad \text{(Equation 9.10)}$$

$$top = Height - bottom = 2 \cdot 4.97 - 3.31 = 6.63 \qquad \text{(Equation 9.11)}$$

Lines pass from "top" through the cut of interest, and from "bottom" through the cut of interest. The new Müller-Breslau points MMB1, MMB2 and MMB3 are established using the previously defined rules shown in Figure 9.10, but with the adjusted angles. The perturbation is shown in Figure 9.28.

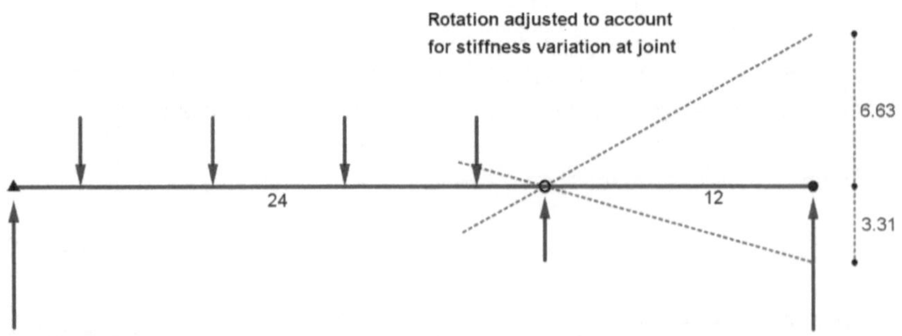

9.28
Correct perturbation that accounts for stiffness variation

Figure 9.29 uses the adjustment necessitated by the stiffness variation, and the perturbation is drawn for the entire beam.

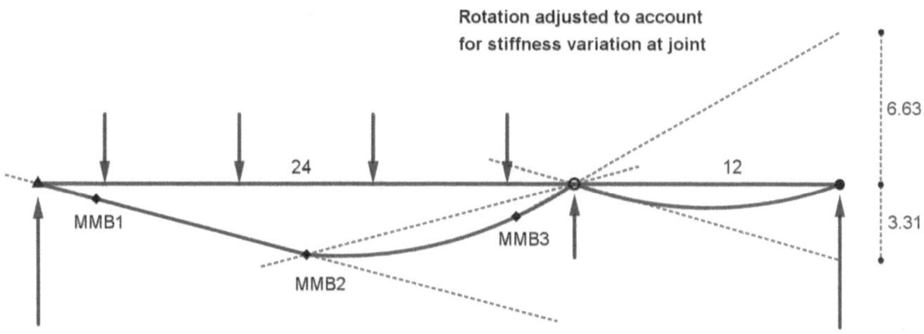

9.29
MB points that account for stiffness variation

Now the problem is quickly solved using Equation 1.1 of Chapter 1, with $\Delta = 1$. The solution is shown in Figure 9.30, with a remarkable agreement to theory.

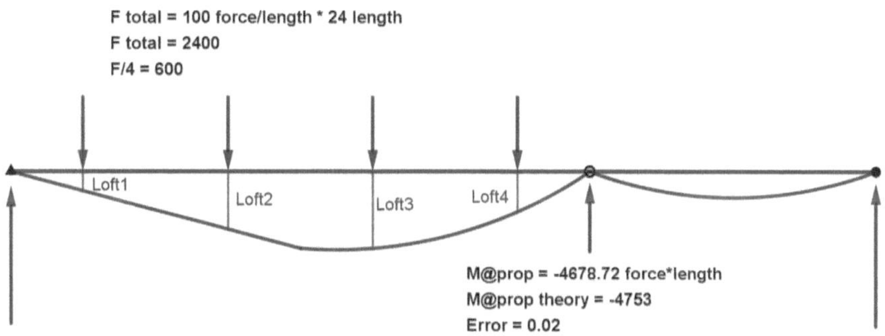

9.30
Solution for the moment in the beam above the interior prop

The use of a simple EI/L definition for stiffness is extremely helpful when studying rectangular frames. By their very nature, frames will often have varying stiffness on either side of the unchanging moment carrying joints at the top of each column.

To develop visual literacy in the reader, the perturbed shape of a single bay, rectangular frame with pinned bases is studied. We seek the horizontal thrust at the lower left support, thus the lower left support slides outward a known Δ. In the following images, Δ will be large enough to easily see the distorted shape of the frame. And note that we must divide by this Δ when solving for the Unknown. Two types of single-bay, single-story pinned base frames will be studied:

The first will have a stiffness ratio of:

$$\frac{\left(\frac{EI}{L}\right)_{beam}}{\left(\frac{EI}{L}\right)_{column}} = \frac{1}{2} \qquad \text{(Equation 9.12)}$$

The second will have a stiffness ratio of:

$$\frac{\left(\frac{EI}{L}\right)_{beam}}{\left(\frac{EI}{L}\right)_{column}} = \frac{1}{3} \qquad \text{(Equation 9.13)}$$

Where E and I are the same symbols as were used for beams, namely E is a designation of the material being used and I describes the cross section of the element. The material property E would never change in such a structure, but the ratio of EI/L for beam versus column could vary because of changes in the length L, or in the cross-sectional property I, or both. Figure 9.31 shows one frame that has a stiffness ratio of 1/2 which arises from cross-sectional property variations as well as geometry. Note the portal opening is square which might imply a ratio of 1. But the reader should note that there are really four variables, I_{beam}, L_{beam}, I_{column}, L_{column}, which reduce to a fraction such as ½. Furthermore, notice the lofts along the left column are different than the lofts along the right column. The reader should sketch these by hand to remember that for a unit horizontal perturbation at the left base support, the loft at mid height on the left column is approximately 0.75 whereas the loft at mid height on the right column is approximately 0.25. The beam mid span loft is never exactly 0.1, but we use approximately 0.1 for ease of sketching. For certain configurations, the central beam loft is 0.2, but the effect of 0.1 versus 0.2 is not enough to justify complicating the suggested lofted shape for sketching. Thus, we use 0.1 for the beam mid span loft, given a horizontal base perturbation of 1.

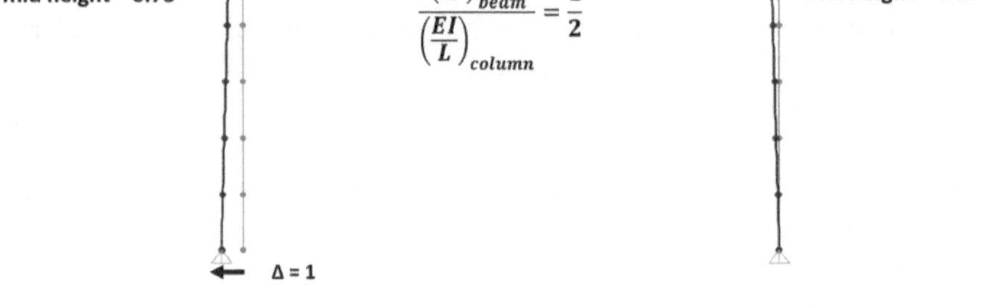

9.31
One setup for a frame stiffness variation of ½

Figure 9.32 shows another frame that has a stiffness ratio of ½ due to the particular I_{beam}, L_{beam}, I_{column}, L_{column} values. Again, notice the lofts along the left column are different than the lofts along the right column.

Note that as always, these perturbed shapes are completely independent of any applied loads. There is no real need to redraw the perturbed shape for the Right Reaction Perturbation, as it is a mirror of the Left Reaction Perturbation. Yet

9.32
A second setup for a frame stiffness variation of ½

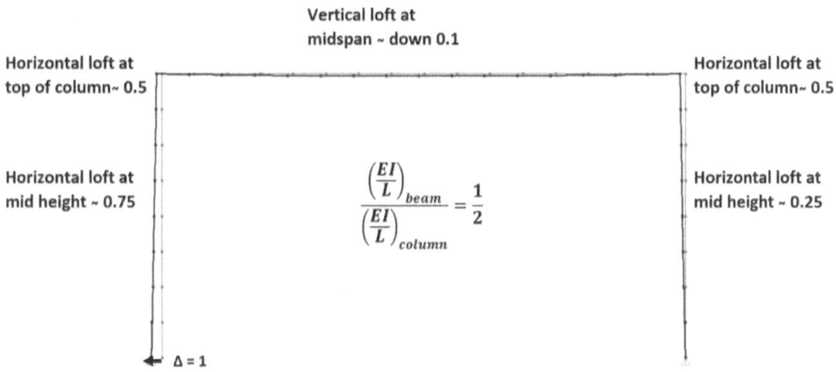

to promote Visual Literacy, here are the shapes for the previous two frames, for horizontal thrust at the bottom right support. Figure 9.33 shows the frame of Figure 9.31 with a perturbation for the lower right equilibrating force. Notice the lofts along the right column are different than the lofts along the left column.

9.33
One setup for a frame stiffness variation of ½, right thrust

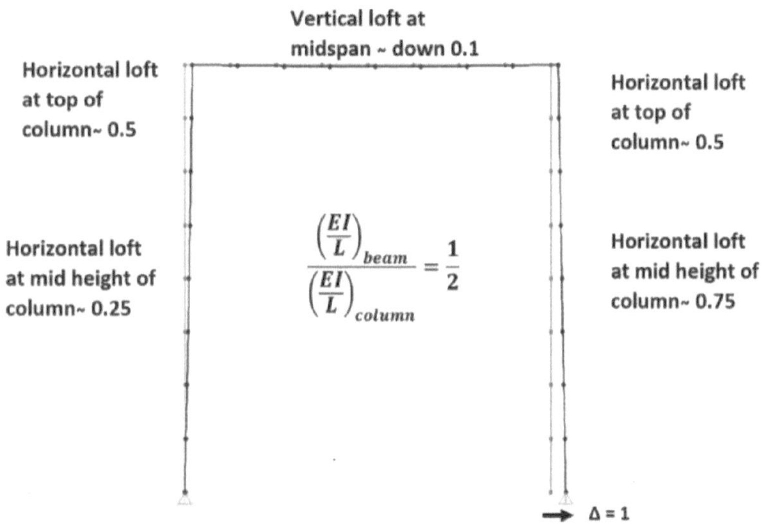

Figure 9.34 shows the frame of Figure 9.32 with a perturbation for the lower right equilibrating force.

Similar results appear for the 1:3 ratio of beam stiffness to column stiffness. Recall that stiffness variations arise from differing cross sections of the elements, or from changes in the lengths of members. Figure 9.35 shows a typical frame with a 1:3 stiffness ratio with a perturbation that seeks the lower left horizontal equilibrating force for any applied loads. As stated before, the beam mid span loft is closer to 0.3 than 0.1 for this shape, but we can choose to simplify the perturbed shape and use 0.1 at beam midspan with little error.

Notice that all of lofts on the beam are small, not only is the midspan loft small. But all of the beam lofts are measurable in Figure 9.35 when compared to the midspan loft. One could reasonably infer what the corresponding loft would be for

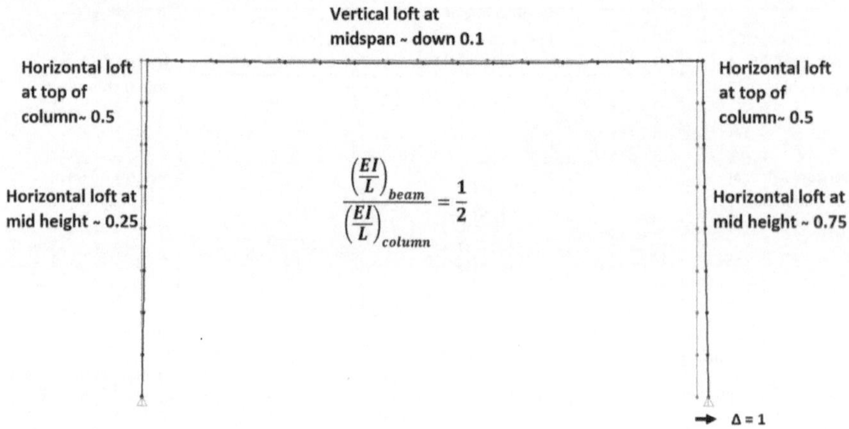

9.34
Another setup for a frame stiffness variation of ½, right thrust

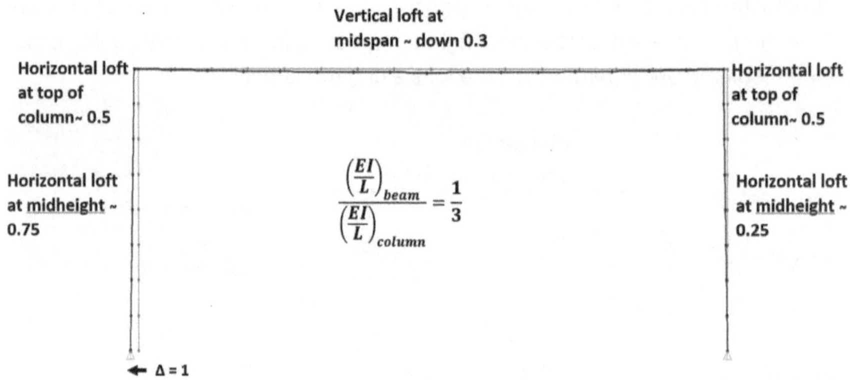

9.35
Perturbation for a frame stiffness variation of ⅓, seeking left thrust

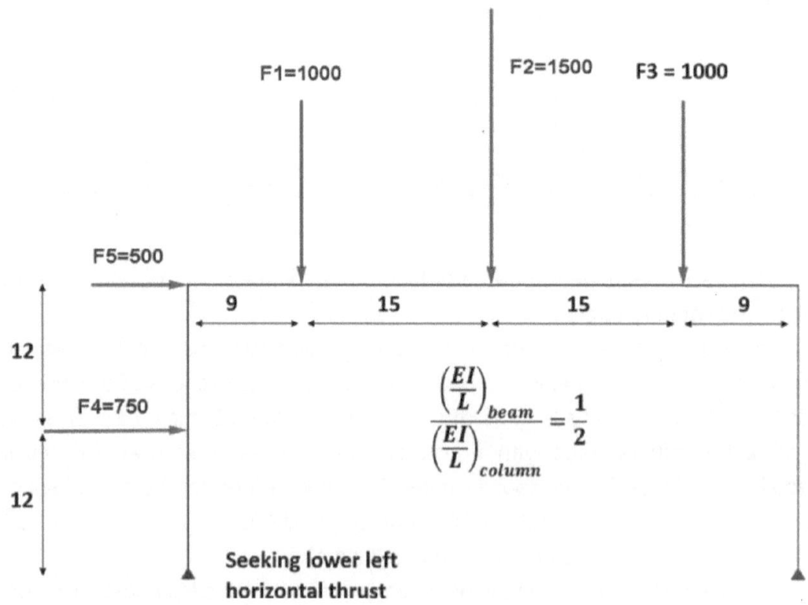

9.36
Example of frame stiffness variation of ½, seeking left thrust

any given load anywhere on the structure, even with a hand sketch. But of course, this technique requires a "visual database" which was just shown. For different configurations, such as a fixed base frame, a new such "database" would have to be generated first in order to use this method. Such a "visual database" helps designers understand how load flows through a structure.

Figure 9.36 shows an example of the 1:2 (beam:column) stiffness frame subjected to five-point loads. The frame has uniform EI in all members, and the height of the frame is 24 units of length tall and the width of the frame is 48 units of length. A GeoGebra model shows the loads at nine units away from the top corners of the frame and at mid-span along the beam. Two lateral loads are shown, one at mid-height and one at the top left corner.

The perturbed shape is drawn in Figure 9.37 as described previously, and lofts are either digitally measured, or manually estimated. Here, they were drawn with GeoGebra.

9.37
Solution finding left thrust

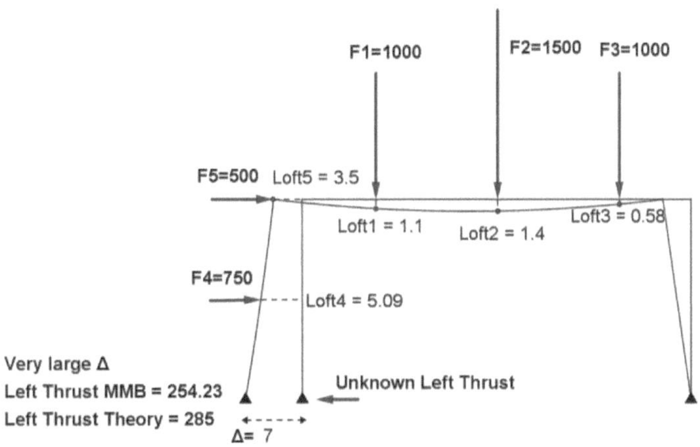

Equation 1.1 of Chapter 1 is used to find the Unknown horizontal left thrust.

$$Unknown \cdot \Delta + F1 \cdot Loft1 + F2 \cdot Loft2 + F3 \cdot Loft3 - F4 \cdot Loft4 - F5 \cdot Loft5 = 0 \quad \text{(Equation 9.14)}$$

Unknown Lower Left Thrust

$$= \frac{1}{\Delta} \cdot (-F1 \cdot Loft1 - F2 \cdot Loft2 - F3 \cdot Loft3 + F4 \cdot Loft4 + F5 \cdot Loft5) \quad \text{(Equation 9.15)}$$

$$Unknown\ Lower\ Left\ Thrust = +254\ units\ of\ force \quad \text{(Equation 9.16)}$$

$$Theoretical\ Lower\ Left\ Thrust = +285\ units\ of\ force \quad \text{(Equation 9.17)}$$

The positive sign means that the thrust is going as assumed, i.e. leftwards. A finite element analysis shows that the thrust is indeed leftwards, with a magnitude of 285 units of force.

Notice in Figure 9.37, that when all the terms are on the left side of the MB equation, the downward loads create positive energy through their lofts, whereas the lateral loads create negative energy through their lofts. Then, when isolating for the unknown, the signs are reversed as shown above.

The analysis continues with a perturbation of the lower right reaction shown in Figure 9.38. This is necessary because of the asymmetry of the problem. Having the lower right reaction will allow for the solution of the moment at the top of the right column, which must match the moment at the right end of the beam if one were to pursue the analysis of the now determinate structure.

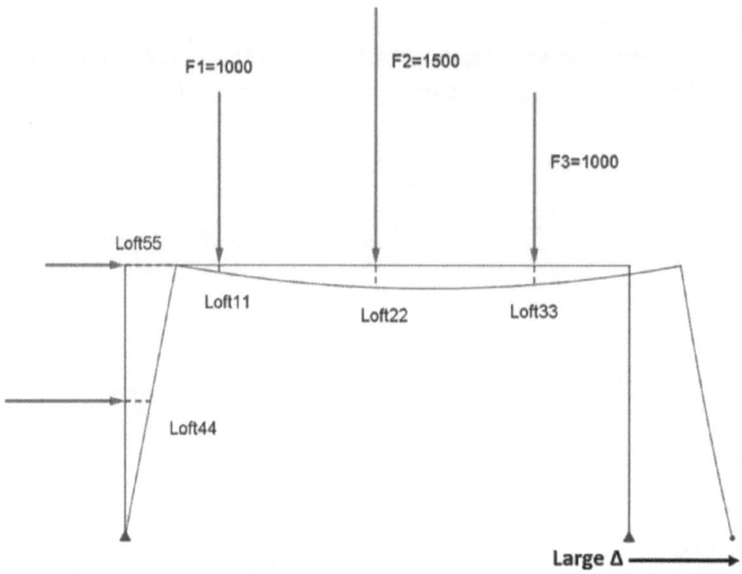

9.38
Solution finding right thrust

Now, the right horizontal reaction is quickly found as:

Unknown Lower Left Thrust

$$= \frac{1}{\Delta} \cdot (-F1 \cdot Loft11 - F2 \cdot Loft22 - F3 \cdot Loft33 - F4 \cdot Loft44 - F5 \cdot Loft54)$$

(Equation 9.18)

Unknown Lower Right Thrust = −975 *units of force* (Equation 9.19)

Theoretical Lower Right Thrust = −965 *units of force* (Equation 9.20)

The negative sign means that the thrust is going opposite of assumed, i.e. leftwards or into the frame. The finite element answer for the right thrust is 965 units of force inward, which nicely matches the MB answer of 975 units of force inward.

Notice in Figure 9.38, that when all the terms are on the left side of the MB equation, the downward loads create positive energy through their lofts, and the lateral loads also create positive energy through their lofts. Then, when isolating for the unknown, the signs are reversed as shown above.

The problem is now fully unlocked. Other moments could be found anywhere on the frame using the previously established techniques.

Even though the MB Method for indeterminate structures is approximate, it may bring insights into load flow because of the visual clarity of its geometric constructions. Consider the frame shown in Figure 9.39, with overhanging cantilevers. We seek the lower left horizontal thrust, but before doing so, we ask ourselves, will the presence of the overhang force FOverh add to, or subtract from any horizontal thrust?

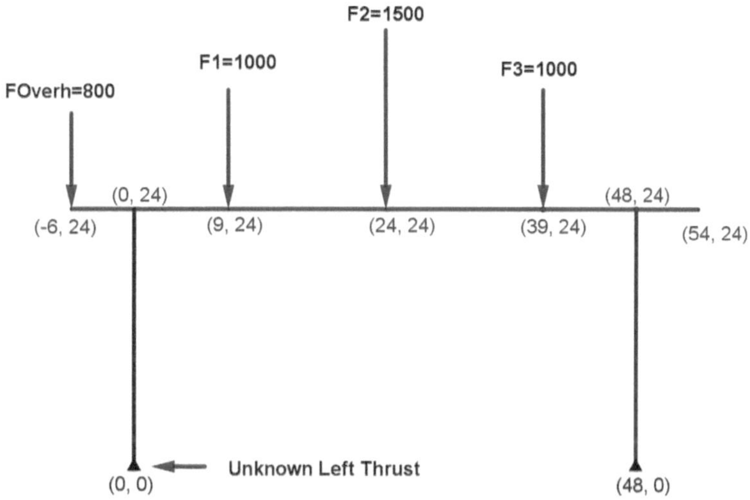

9.39
Indeterminate frame with overhangs, one of which is loaded

The work created by F1 F2 and F3 in Figure 9.39 is positive, whereas the work of FOverh is negative. Thus, the interior forces create an inward thrust, which would be negative by the assumption of Δ in Figure 9.38, whereas the overhang force would create outward thrust. The solution for the lower left horizontal equilibrating reaction is shown in Figure 9.40.

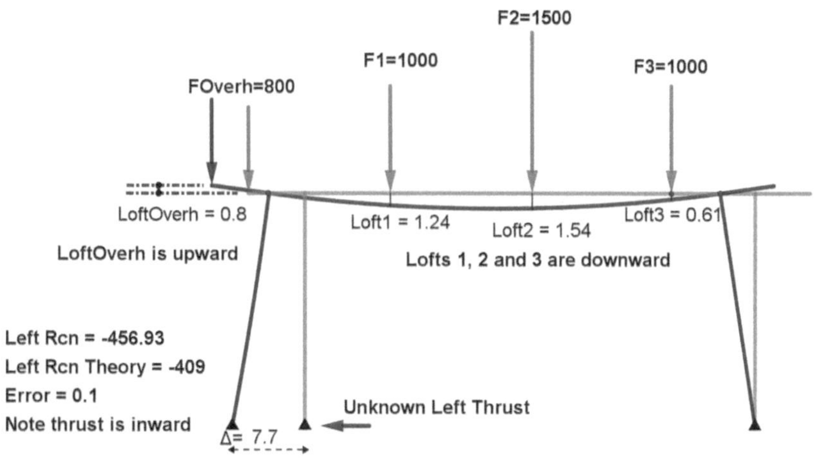

9.40
Solution finding left thrust

INDETERMINATE STRUCTURES ÉTUDES

The following studies or études should be sketched by hand.

Étude 9.1) An indeterminate beam is 12 units long. It is fixed at the left end and is propped at a point 10 units of length from the left, fixed end, leaving a 2 unit overhang. The entire beam is subject to a uniform downward load of 100 force/length. Find the equilibrating moment in the left fixed end. Consider breaking up the uniformly distributed load into two big sections, one from the wall to the prop, and one from the prop to the free end. Then break up the first part of the load into at least four parts, and break up the second part of the load into at least two parts. A hand sketched solution is shown in Figure 9.41. The "theory" solution is based on a typical structural beam stiffness and is meant to be a guideline of how close the answer is to known solutions.

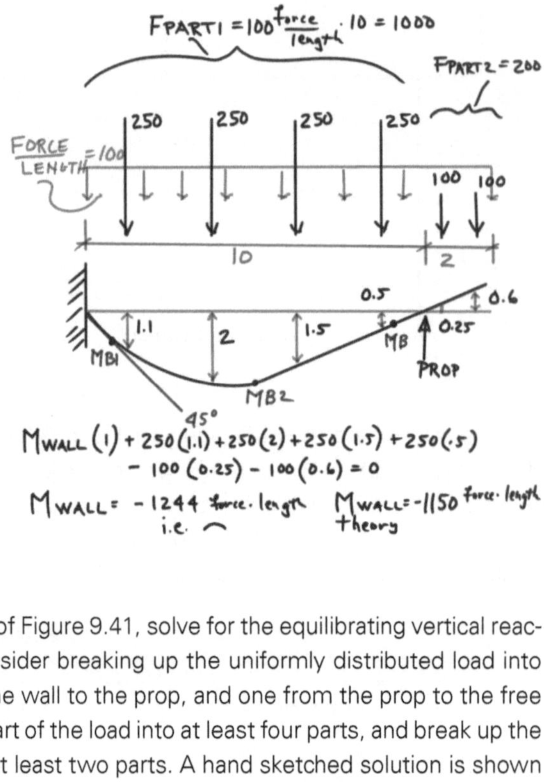

9.41
Solution for Mwall in a propped cantilever, étude 9.1

Étude 9.2) For the beam of Figure 9.41, solve for the equilibrating vertical reaction in the prop support. Consider breaking up the uniformly distributed load into two big sections, one from the wall to the prop, and one from the prop to the free end. Then break up the first part of the load into at least four parts, and break up the second part of the load into at least two parts. A hand sketched solution is shown in Figure 9.42. The "theory" solution is based on a typical structural beam stiffness and is meant to be a guideline of how close the answer is to known solutions for such an indeterminate problem.

Étude 9.3) For the beam described in Étude 9.1, suppose we were given the equilibrating moment in the wall as −1150 force*length. Given that moment as well as the applied distributed load, find the equilibrating vertical force at the prop support. Recall from Chapter 2 that we studied a concentrated moment at the end of a beam specifically for this reason. Review Figures 2.43 and 2.44 before solving this étude. A hand sketched solution is shown in Figure 9.43. Note that having the moment in the wall makes this a determinate problem, not an indeterminate one.

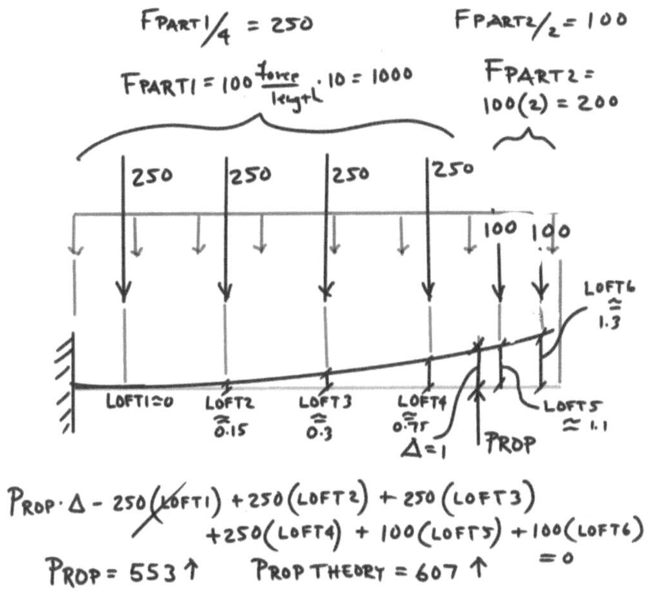

9.42
Solution for the equilibrating prop force, indeterminate beam, étude 9.2

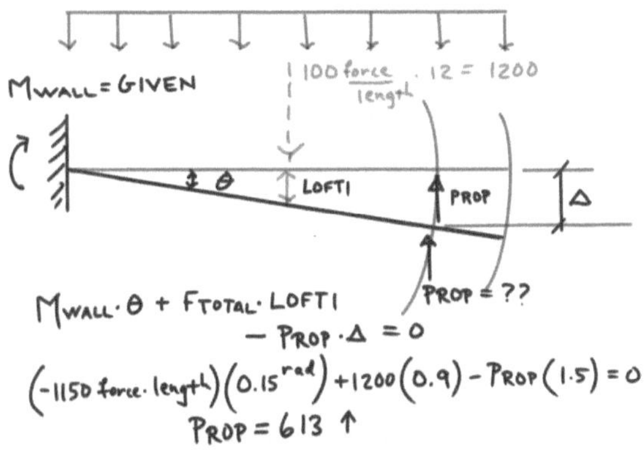

9.43
Solution for the force in prop given moment, determinate problem, étude 9.3

Also note that the theoretical prop equilibrating force was given in Figure 9.42 as 607 units of force upwards.

Étude 9.4) A beam is 24 units long. It is pin supported at the left end, roller supported at midspan, and at the right end. A uniformly distributed load of 150 force/length is applied to the central half of the span, i.e. starting at 6 units of length from the left end and ending at 18 units of length from the left end. Find the magnitude of the central equilibrating prop support. You may invoke the powerful argument of symmetry, but break up the uniform load into at least four parts on either side of the prop.

Before you start, estimate what percentage of the total load will flow into the central prop. 50%? 75%? Were you surprised by the answer? A hand sketched solution is shown in Figure 9.44.

Étude 9.5) For the beam of étude 9.5, find the moment in the beam directly about the central prop. A hand sketched solution is shown in Figure 9.45.

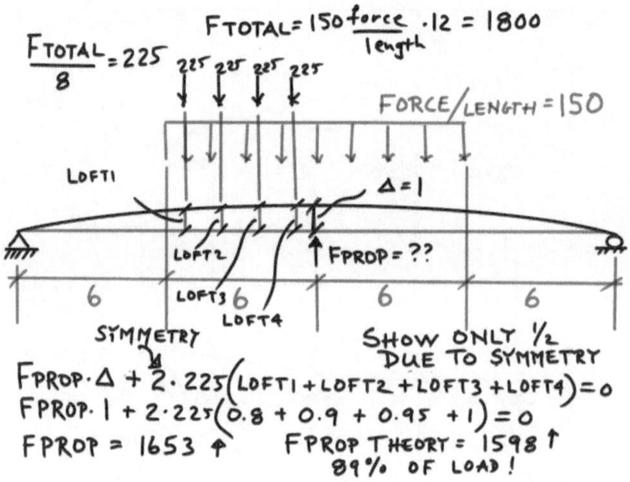

9.44
Solution for the force in prop, étude 9.4

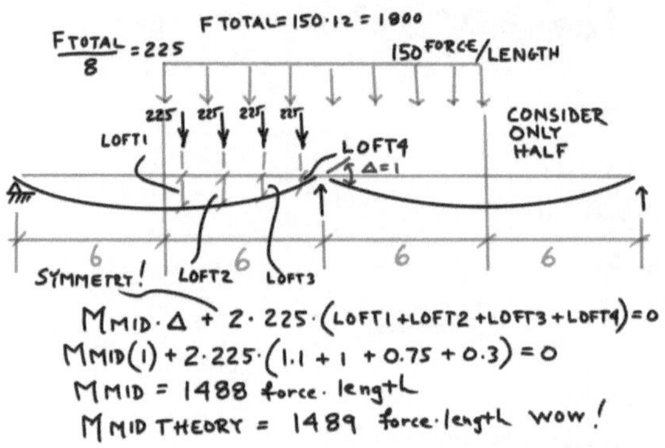

9.45
Solution for the moment in the beam directly above prop, étude 9.5

Étude 9.6) An indeterminate frame is 12 units tall and 24 units wide. The cross-sectional properties and the material are the same for the beam as for the columns. For a uniformly distributed load of 100 force/length applied on the beam, find each of the lower two equilibrating horizontal thrusts at the bases. Note that only one perturbed shape is truly needed, it can be flipped, or the powerful argument of symmetry can be invoked. A hand sketched solution for the lower left thrust is shown in Figure 9.46. A hand sketched solution for the lower right thrust is shown in Figure 9.47. Be explicit with the signs of the thrust, i.e. show clearly if the thrust going to the right or to the left for each solution.

Étude 9.7) Repeat étude 9.6, but this time there is only a single point load of 1,000 units of force applied horizontally at the top of the left column, with no distributed load. Find each of the lower two equilibrating horizontal thrusts at the bases. Note that two perturbed shapes are needed, because of asymmetry. A hand sketched solution for the lower left thrust is shown on the left of Figure 9.48 and a hand sketched solution for the lower right thrust is shown on the right of Figure 9.48. Be explicit with the signs of the thrust, i.e. show clearly whether the thrust is going to the right or to the left for each solution.

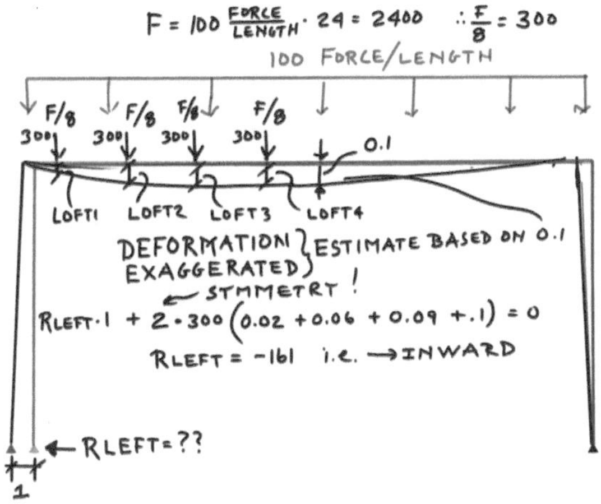

9.46
Solution for the lower left thrust due to distributed load, étude 9.6

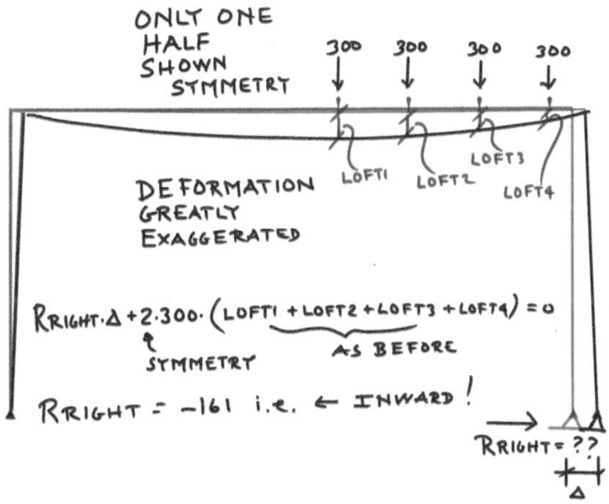

9.47
Solution for the lower right thrust due to distributed load, étude 9.6

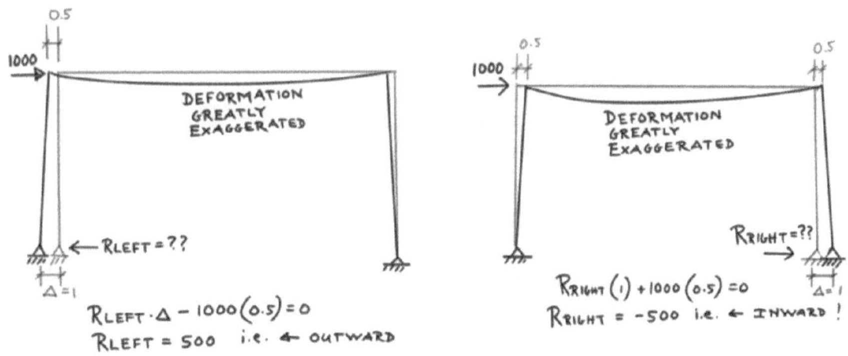

9.48
Solution for both thrusts due to point load, étude 9.7

Étude 9.8) Repeat étude 9.6, but this time apply both the uniformly distributed load, as well as the horizontal point load. Check to see that the answers of étude 9.6 and étude 9.7 superpose to give you the answer for étude 9.8, but actually solve étude 9.8 for both loads. Be explicit with the signs of the thrust, i.e. show clearly whether the thrust is going to the right or to the left for each solution.

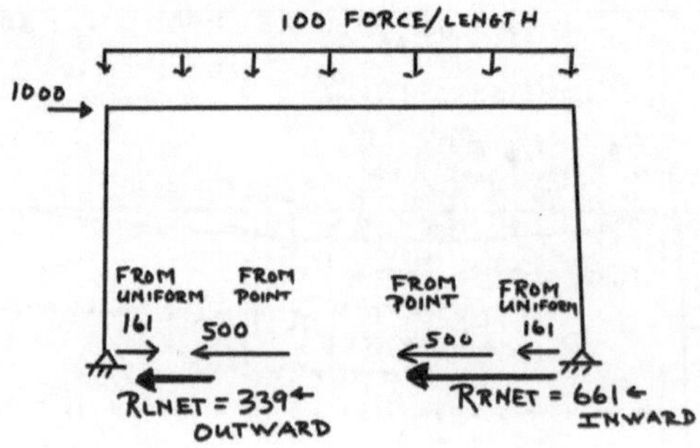

9.49
Solution for the superposition of both loads, étude 9.8

A hand sketched solution for the superposition of both the distributed load case as well as the point load case is shown in Figure 9.49. Clearly superposition of étude 9.6 and étude 9.7 is valid.

10 Special Topics

This final chapter introduces two special topics that are traditionally considered advanced, but they can immediately be solved by novice readers via the geometric tools we have been using. These two classes of problems demonstrate the power of the Modern Müller-Breslau Method, and hopefully, they will intrigue the reader enough to pursue these and other related topics via geometric solutions rather than algebraic ones. A few solutions for complicated buckling problems will be solved quickly via the Modern Müller-Breslau Method. Then a few solutions for plastic analysis of beams will be shown. Traditionally, such problems are the domain of graduate-level inquiry, but here we sweep aside all theory and simply use Equation 1.1 of Chapter 1 in a slightly modified format, to solve the problem of what load induces a collapse, or instability, known as buckling. No major theoretical ideas will be presented as background to these two topics, yet the theoretical answers will be given and readers who are curious are encouraged to dig more deeply into each of these topics.

We begin by briefly studying buckling or instability examples. Buckling is a complicated and dangerous phenomenon because it is the sudden collapse of a compressed element, which seemingly was safe, but it instantly changed shape with an increase of load, creating instability. The mathematics generally includes differential equations and complex theorems, yet all this is brushed aside when viewed from a Modern Müller-Breslau point of view. As the perturbation gets small, the Modern Müller-Breslau answer approaches the theoretical answer.

Consider the following images to rapidly get acquainted with the classical instability or buckling problem. A perfectly straight column is pinned at its base, free at its top, and is subjected to a perfectly straight downward compressive load at its top. In such a perfect world, there is nothing to fear, the column can support the load. But nothing is perfect, and if the gentlest breeze moved the column a bit left or a bit right, it would suddenly become unstable as shown in Figure 10.1. It is this phenomenon that we are concerned with, a sudden change in geometry that leads to a collapse. It is a very dangerous phenomenon indeed, as one could well imagine trying to hold a large member simply by pinching it at its base as shown in Figure 10.1. The reader is encouraged to pinch a pencil only on its sharp lead point and then to test how such a pinching restraint is unable to prevent rotations of the pencil.

Next, suppose a lateral spring was placed somewhere near the top of the column in Figure 10.1. The spring has stiffness force/length and when the column

DOI: 10.4324/9781003180913-10

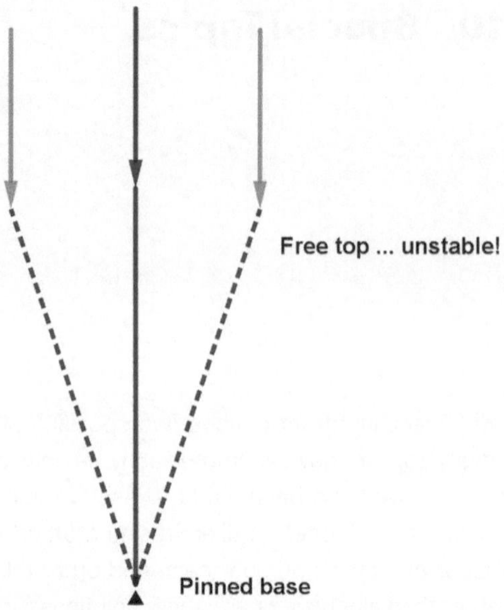

10.1
First buckling example setup

is vertical, the spring is unstretched; it has no force in it. Perhaps it is another structural element, such as a floor of a building, but that floor must not be able to drift freely; it must be anchored at the end away from the juncture with the column. Now, the structure again seems to be stable and safe, as shown in Figure 10.2. But danger still lurks!

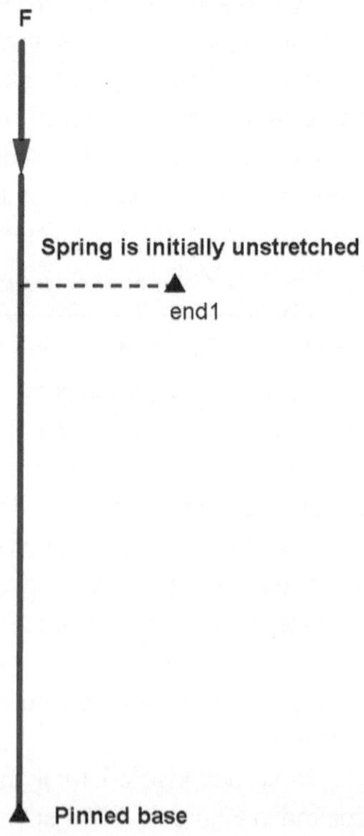

10.2
Attempting to stabilize column

For a finite rotation of the column, or using the language of this textbook, for a small perturbation of the original geometry, one can intuitively feel that a force develops in the spring as the spring stretches, or compresses, to accommodate the rotation of the column. This is shown in Figure 10.3. Note that even though the stretched spring does not appear to be horizontal, the assumption is that loads always maintain their original orientation, thus the spring force would still be horizontal in the perturbed condition.

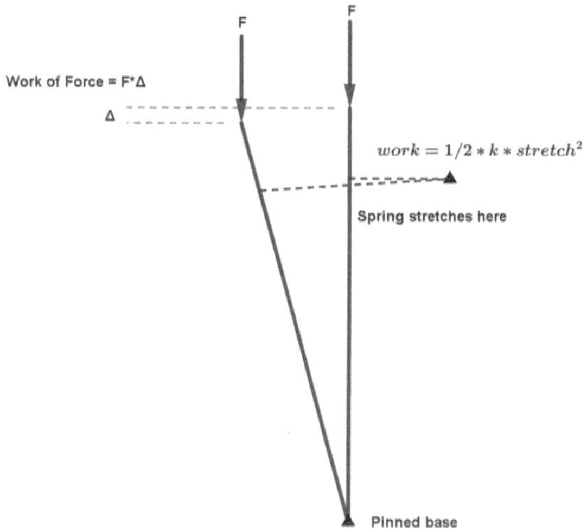

10.3
Force slowly develops in the string during perturbation

Hopefully, it is intuitive that for a very flexible spring, i.e. one that provides barely any force when stretched, the assistance of the force developed by such a flimsy spring would be next to nothing and the column still collapses. For a more robust and stiff spring, larger and larger perturbations could be safely handled prior to a collapse. It is this collapse load we seek, for a given spring and a given geometry. Notice that the work done by the vertical applied compressive force F is:

$$\textit{Work of External Force} = \textit{Force} \cdot \Delta \qquad \text{(Equation 10.1)}$$

This is so because the vertical load is already on the column and it simply moves through a vertical Δ. However, the force in the spring slowly develops. In the original unperturbed configuration, we assume that all springs are unstretched, i.e. springs initially have zero deformation. In this original configuration, the force in any spring is zero, yet the force slowly climbs up as the spring begins to stretch (or compress). Thus:

$$\textit{Work of Spring} = -\frac{1}{2} \cdot \textit{Force} \cdot \delta = -\frac{1}{2} \cdot (k \cdot \delta) \cdot \delta = -\frac{1}{2} \cdot k \cdot \delta^2 \qquad \text{(Equation 10.2)}$$

Where k is the given spring stiffness (force/length) and δ is the stretch of the spring. Note that the force in the spring is $k \cdot \delta$.

It may be helpful to think of work as the area under the force versus displacement curve. Figure 10.4 shows this graphically. The applied compressive force F never changes through the perturbation, and it is constant, whereas the Spring

Force is solely dependent on the displacement; the Spring Force increases only when the displacement increases. The Spring Force starts at zero and linearly builds up as the spring elongates or compresses. Thus, for work where the load is always in place and simply rides through a displacement, the work is "rectangular", as on the left of Figure 10.4, whereas if the load slowly builds up from zero, the work is "triangular", as on the right of Figure 10.4.

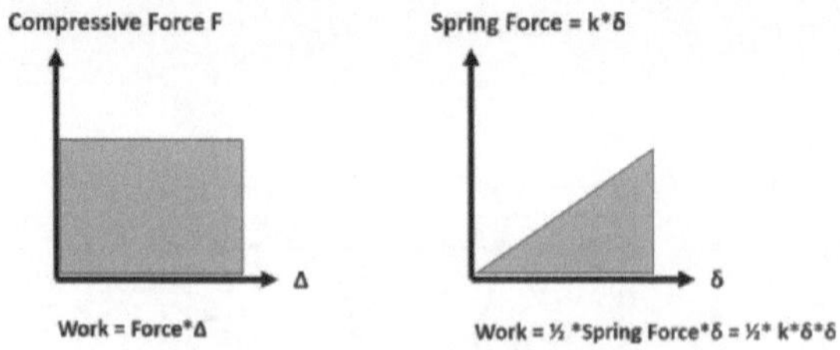

10.4
Two definitions of work

Using the insights of Figure 10.4, it is necessary to slightly modify Equation 1.1 of Chapter 1 as follows. Notice that the sign of the work done by the spring is always negative. If the spring stretches, the tension in the spring resists the stretch, if the spring shortens, the compression in the spring opposes the shortening. In other words, the spring force and the spring deformation are always in opposition, creating negative work. The modified equation we will use for such buckling problems is:

$$Force \cdot \Delta - \frac{1}{2} \cdot k \cdot \delta^2 = 0 \qquad \text{(Equation 10.3)}$$

Finally, note that in the modified Equation, there is still only one unknown, as the geometric constructions of rigid structural elements will always link Δ, the movement of the axial compressive force, to δ, the deformation of any spring. Figures 10.5 and 10.8 were originally studied by W.F. Chen and E. Lui, in their study of Stability Design of Steel Frames. The problems have been solved in a novel manner here.

Figure 10.5 describes a typical academic problem studied by advanced students. Two rigid elements are hinge connected to each other, but laterally restrained by flexible elements known as springs. There is also an inflexible roller support at mid-length and an immovable pin support at the left end. The question is this, given some spring stiffness k for each spring, whose units are force/length, what load P critical would cause the entire structure to snap into an unstable configuration and consequently collapse?

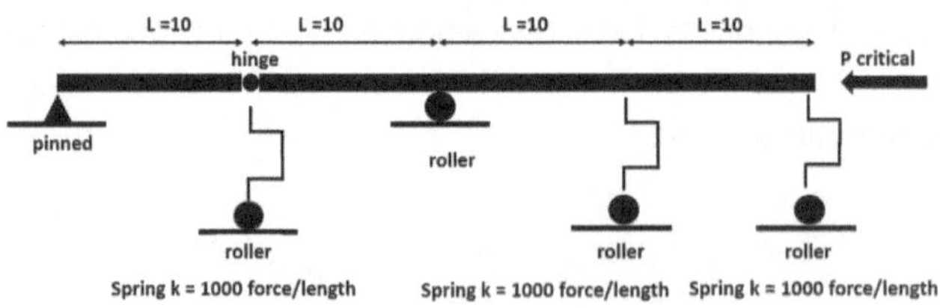

10.5
First buckling example setup

The solution for this complicated problem is exactly the same as the solution for any other problem in this textbook, namely apply the modified Equation 1.1, and solve for the Unknown, here the Unknown is Pcritical, i.e. the buckling or instability load. The quick geometric solution is shown in Figure 10.6.

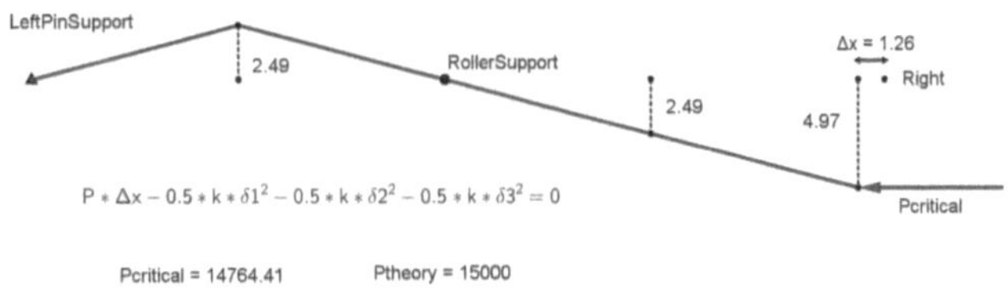

10.6
Estimate of Pcritical for large Δ

As was seen in other problems that used the Modern Müller-Breslau Method, as Δ is made small, the answer approaches the theoretical value. Figure 10.7 shows the solution for a small Δ.

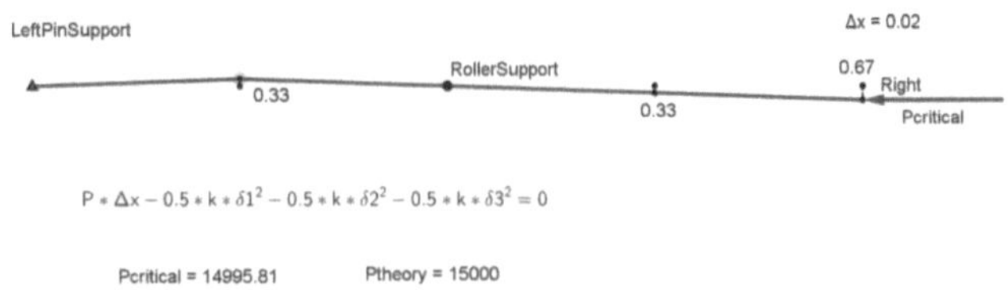

10.7
Estimate of Pcritical for small Δ

The same technique is applied to yet another academic stability or buckling problem which is shown in Figure 10.8.

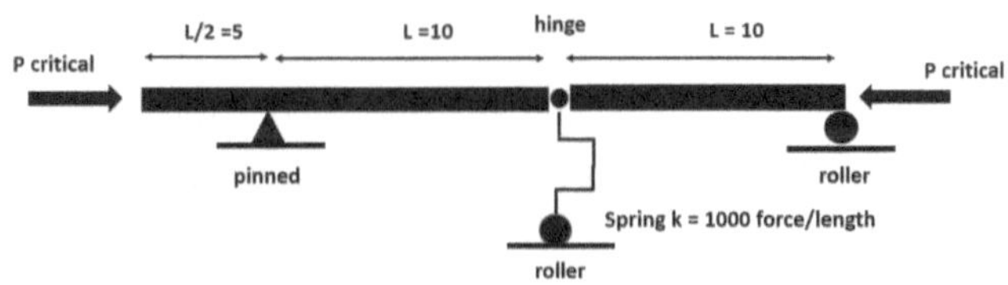

10.8
Setup for a second buckling example

Figure 10.9 shows the solution for Pcritical, the buckling load, using a large Δ Right. Here, the Δ on the Right is not only geometrically linked to the δ deformation of the spring, it is also linked to the Δ on the Left. The reader need not worry about sines and cosines when seeking these perturbations, they can simply be measured on the drawing.

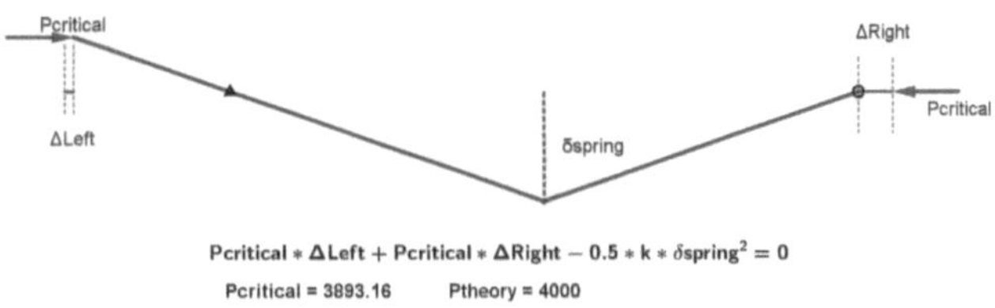

10.9
Solution for large Δ Right

Figure 10.10 shows the solution for a small ΔRight.

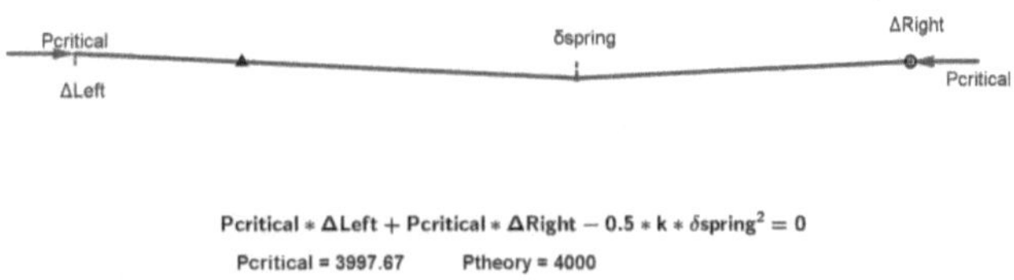

10.10
Solution for small Δ Right

The second topic we explore in this final chapter is also traditionally studied by advanced students and by practitioners, namely the study of plastic hinge formations in structures, and the determination of what load would cause such a collapse mechanism to form. As loads are slowly increased at a fixed location on a structure, the demand required of the structure becomes greater and greater. Eventually, local bending failure can occur. As an analogy to help the novice reader understand this idea, a few easy to understand images are discussed focusing on timber members which fail quickly. Such sudden, brittle failures are not what happens in ductile steel structures. We use plastic analysis in steel structures, not timber ones. Yet, to understand the concept, consider familiar examples. One common everyday experience of a hinge forming in a structure may be too much snow on a cantilevered tree branch, which suddenly snaps near the trunk, at the point of maximum bending. At this juncture, the branch cannot transfer bending to the trunk anymore, but the branch itself may still be intact elsewhere along its length. Another brittle failure example is an attempt to cross a stream with a flimsy board, and just when the person is midspan on the makeshift "bridge," the board snaps, again at the point of maximum bending, directly under the person's feet. The two remaining pieces of the board may be intact, but the hinge formed at midspan has caused a collapse failure to occur. These two common examples both began with determinate structures, namely a cantilevered branch and a simply supported board resting on either side of a stream. For determinate structures, one "cracked hinge" point is sufficient to cause an instability failure. In fact, the curious reader may see such potential instabilities in the previous buckling examples of Figures 10.5 and 10.8. Yet, in Figures 10.5 and 10.8, the location of the hinge was given to us. In plastic analysis, we ***do not know*** where the "cracked hinge" will form! But it makes the most logical sense to guess that the "cracked hinge" will form at the point of maximum bending, for example at the connection of branch to trunk in our cantilever example, or at midspan directly under the person's feet in the board crossing the stream example.

Because of the lack of perfect clarity of where a hinge will form, and because the method often overpredicted failure loads, the method fell out of favor from its heyday in the 1970s. But a renaissance of sorts is underway for plastic analysis today, because of the so-called "pushover" method, wherein a structure is laterally loaded with ever-increasing loads, and the sequence of collapse mechanisms are studied in great detail, via sophisticated computer models. Yet once again, in this textbook we brush away all mathematical complexities and solve the plastic analysis problem solely geometrically with the Modern Müller-Breslau Method.

A basic premise of plastic analysis is that a hinge is formed when a moment exceeds a "plastic" level known as MP which is an acronym for "Moment Plastic." For steel, such an MP is well defined, it is based on the material properties of the element, as well as the cross section of the element being bent. MP is always known before the problem begins, i.e. it is a constant value that is not dependent on any loading, it is simply a property of the element being bent. This concept is one of the more challenging ones to simply accept, namely that MP is given before the problem begins, as it is based solely on the material properties and the cross section of the element undergoing bending. Once the bent element feels MP, an internal hinge is formed at that point. Thus, the goal of plastic analysis is ***not*** to find MP, but rather to find the least load which causes the collapse of the entire structure due to the formation of these internal hinges.

Another basic premise of plastic analysis is that the internal moment can never exceed MP, it is the largest conceivable moment that the element can experience. Yet another basic premise of all the analyses we have conducted so far is that moment causes rotation, and doubling the moment doubles the rotation, tripling the moment triples the rotation, etc. But at MP, the moment can no longer increase, yet the rotation can freely continue to grow, as a perfect hinge has now been formed at that point where MP occurred. In other words, the structure's support conditions literally change once a hinge is formed. A qualitative example using a propped beam will clarify this concept. Figure 10.11 shows a propped cantilever with a single force applied at midspan.

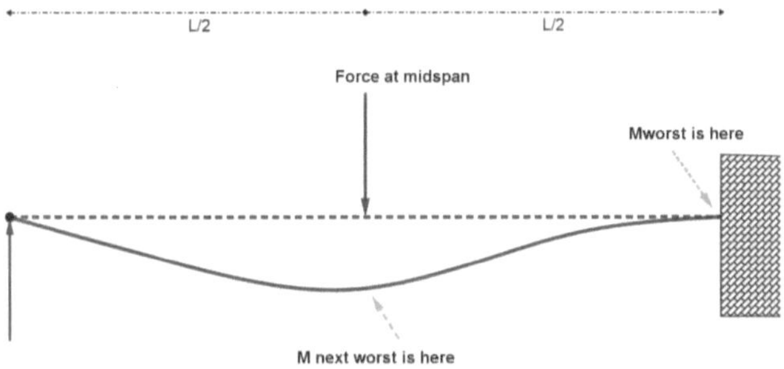

10.11
Indeterminate beam subjected to one load at midspan

It may not at all be obvious to either the novice or to the advanced reader which moment is larger, the moment at the right wall, or the moment under the concentrated load. Certainly, the techniques of indeterminate MB analysis could have been used to study each cut individually, but that is time consuming and a bit fussy. A better technique is to intuit the curvature of the beam and to note that where the curvature, ***not the deflection***, is greatest, there will be the greatest bending moment. In a determinate structure, we place one hinge at the point of maximum curvature which corresponds to the point where the maximum moment will occur. In an indeterminate structure, we keep repeating this process until a "mechanism" or instability exists. The order of appearance of hinges does not really matter, but the placement of the hinges does matter.

To continue the qualitative example begun in Figure 10.11, we intuit that the moment will be worst either at the right wall or directly under the load. We intuit this by a "visual database" of how such a beam would curve under given loads, a topic that is beyond the scope of this textbook. Yet, since the order of hinge appearance does not really matter for us in this textbook, we assume that the point of contact with the wall was the worst. Notice how curious the changes from Figures 10.11 to 10.12 are! In Figure 10.11, the moment was maximum in the right wall as it furiously restrained the beam to ensure zero slope. As the applied force continues to increase, that equilibrating wall moment eventually peaks at MP. Suddenly in Figure 10.12, everything has changed and the shape of the deflected beam in Figure 10.12 demonstrates that a new structure has been created, one that is wildly different than the previous one. The structure in Figure 10.12 is still stable, however, but now it is a classic pin-ended determinate beam and the worst additional moment must be directly under the point load at midspan. The load increases a bit more since the beam is already bent severely, and the final hinge is formed. This is the essence of the "pushover analysis" previously mentioned. Hinges sequentially form as the structure continues to deteriorate under ever-increasing loads.

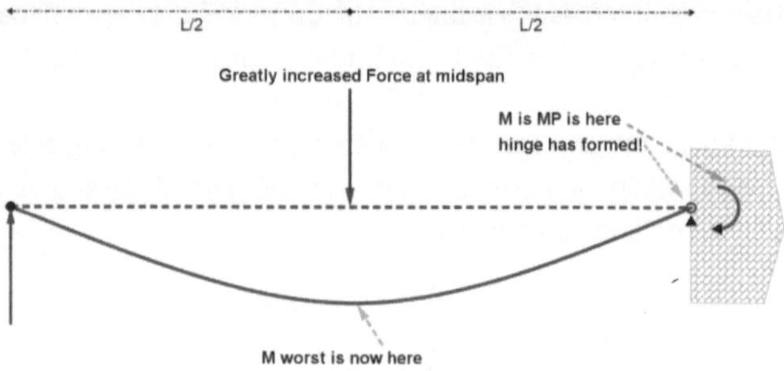

10.12
Suddenly determinate beam subjected to one load at midspan

Figure 10.13 shows the second, and in this case, the final hinge that is formed, making the structure unstable and collapse ensues. What may puzzle the novice reader is the fact that MP, the "Moment Plastic" is actually present at these hinges. It is what caused the hinge to form. The moment is not zero at these hinges!

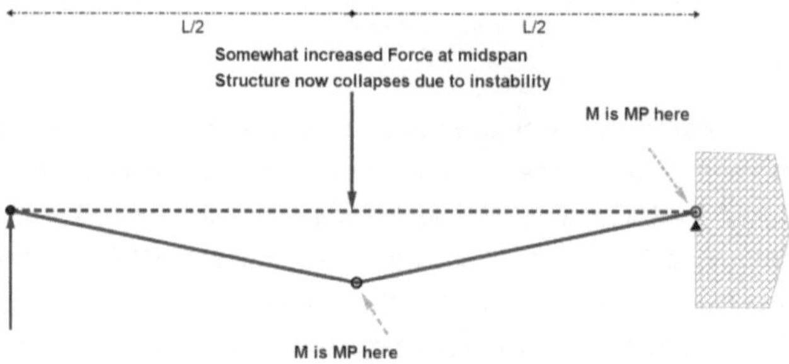

10.13
Suddenly unstable beam subjected to one load at midspan

The ultimate question we want to answer with plastic analysis is this:

- What force will cause enough hinges to form due to MP being reached in one or more places, such that the structure collapses?

We will use Equation 1.1 of Chapter 1 yet again, but this time, the Unknown is the force being applied in ever-increasing value. As was done in Equation 4.3 of Chapter 4, the internal work is negative since the induced moment opposes a forced rotation. Consider Figures 10.14 and 10.15. If the beam is cracked and forced to perturb upward by a constraint such that the rotation is positive, the induced moments are negative, i.e. "sad" or tension on the top caused by that perturbation. Note, this is the moment induced by a perturbation, not rotation induced by a moment. This is shown in Figure 10.14. Yet, if the beam is cracked, and forced to perturb such that the rotation is negative, the induced moments are positive, i.e. "happy" or tension on the bottom. This is shown in Figure 10.15.

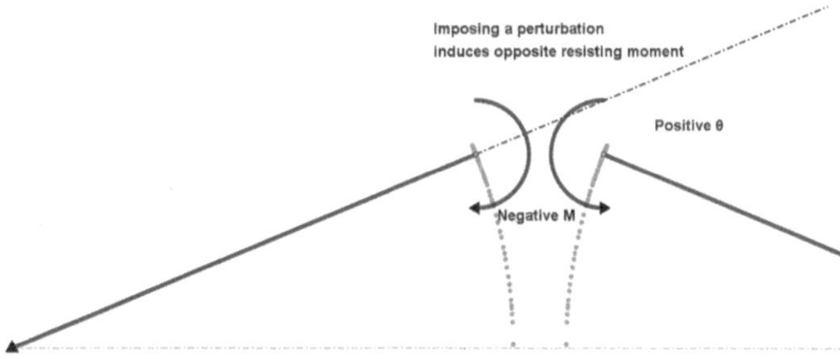

10.14
Sign convention, work of the internal moment negative one way

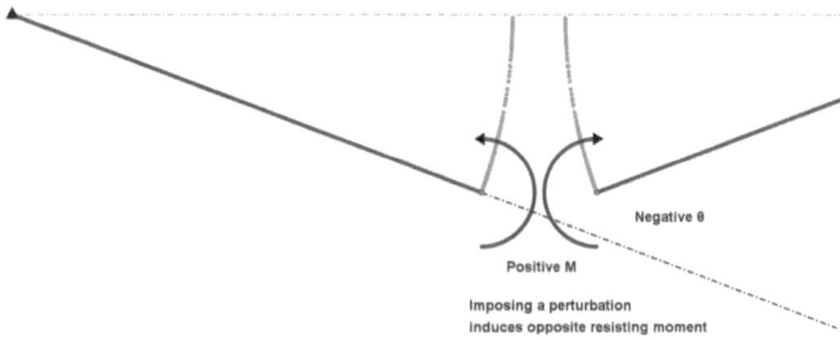

10.15
Sign convention, work of the internal moment negative the other way

Thus, Equation 1.1 of Chapter 1 is slightly rewritten, just as it was for internal work in trusses. Equation 10.4 shows this idea for Plastic Analysis in conjunction with the Modern Müller-Breslau Method.

$$\sum -Moment_i \cdot \theta_i + \sum Force_i \cdot loft_i = 0 \qquad \text{(Equation 10.4)}$$

To demonstrate one way of solving this problem via a parametric drawing environment, a propped cantilever, fixed at the left end, was drawn in GeoGebra and shown in Figure 10.16. The cut point where the moment is sought is a variable, and the placement point of the concentrated load is also a variable. We know the theoretical value of the minimum F which causes a collapse. But we set up a somewhat general drawing first, and then we will move the cut point and the load application point until we visually establish a minimum force. Figure 10.16 shows a solution that is not a minimum since the F found is 8.49 times the known constants MP/L where L is the Span of the beam.

As the cut point is manipulated and the Load application point is independently manipulated, the answer is instantly displayed, telling the user that Fmin is changing. A non-optimized Fmin is shown in Figure 10.16.

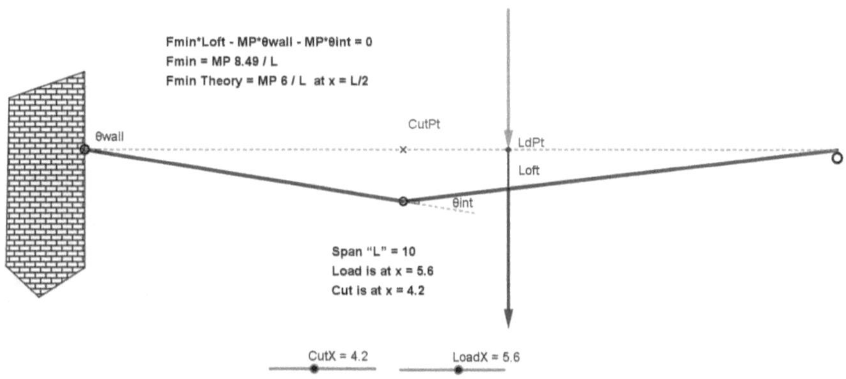

10.16
Exploring possible Fmin for a propped cantilever beam subjected to one load anywhere

Figure 10.17 shows further explorations of Fmin. This demonstrates that no previous knowledge of Fmin is really needed, one can track the effect of moving the cut point or the load point.

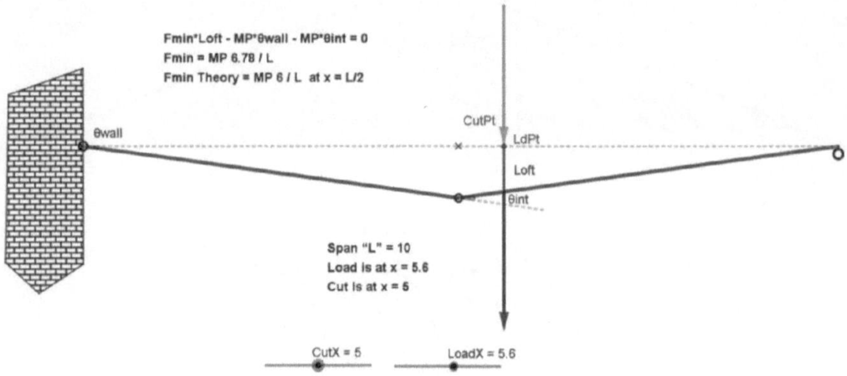

10.17
Smaller Fmin for a propped cantilever beam subjected to one load anywhere

After further manipulations, one quickly establishes that the minimum Fmin occurs when the cut point is placed at L/2 and the load application point is also placed at L/2. Then, the Fmin value approaches the theoretical value, as the angle is decreased. An optimum solution is shown in Figure 10.18. Note that no knowledge of the theoretical minimum value was needed to solve this problem, and no such knowledge was expected. We simply move the cut and track the effect of the movement, seeking a minimum failure load. What was needed however was the knowledge that two hinges must form in this beam, and that one hinge is at the left wall.

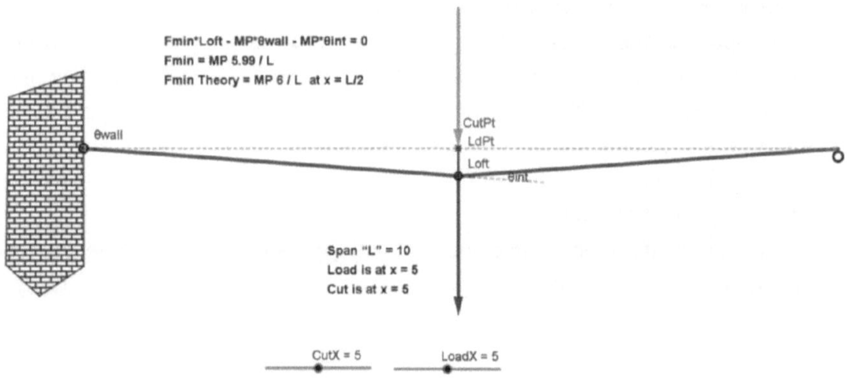

10.18
Absolute minimum Fmin for a propped cantilever beam subjected to one load anywhere

The power of the method is demonstrated with slightly more complicated examples. In the previous situation shown in Figure 10.16, it may have been intuitive to manually place hinges at the wall support and directly under the load. But in the following example, a propped cantilever is shown with two identical point loads, each load is applied at the one-third point along the span and the loads do not move. Note that each load has the same value F1, and we seek the minimum value of F1 that causes a collapse. Notice that there is only one Unknown in Equation 1.1 of Chapter 1, and that unknown is F1min. We know that we need two hinges to form in order for the structure to collapse, we assume that one hinge is at the wall, but where might the second hinge form? The problem setup and an initial solution is shown in Figure 10.19.

10.19
Propped cantilever with F1 at L/3 and same F1 at 2L/3

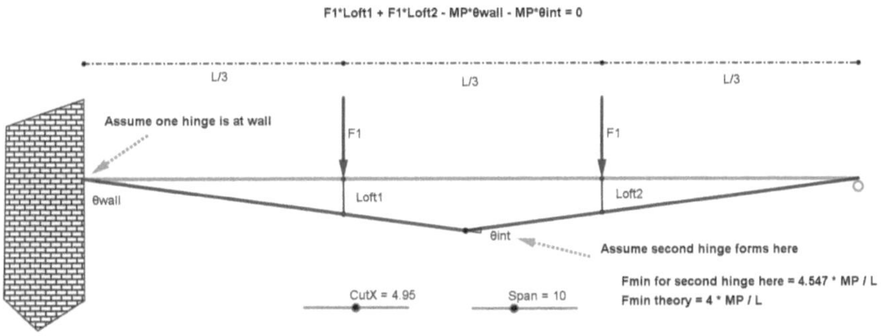

Manually moving the cut point for the second hinge to be directly under the load at L/3 gives an F1 minimum value of 5 MP/L, as shown in Figure 10.20. Yet, the theoretical minimum F1 value is 4 MP/L. Even if we didn't know the theoretical value, which the novice reader certainly will not know, we can manually move the cut point along the length of the beam and simply watch what value of F1 is found from Equation 1.1. When F1 is minimized, we have the most dangerous mode of failure with the smallest possible load to cause such a failure.

10.20
Manually placing hinge at L/3 gives Fmin = 5 MP/L, not optimal

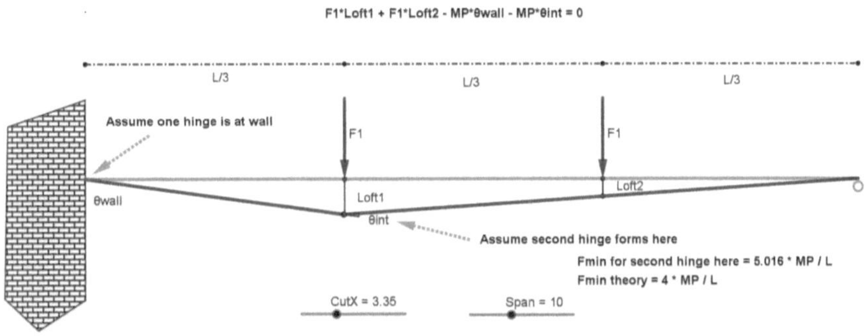

Further manual manipulation of the second hinge location shows a minimum of 4 MP/L when it is placed directly under the load at 2L/3. Sliding the cut further towards the right prop support gives ever-larger F1 values, thus we deduce that Fmin = 4 MP/L when the second hinge is at 2L/3. This agrees with theory as shown in Figure 10.21.

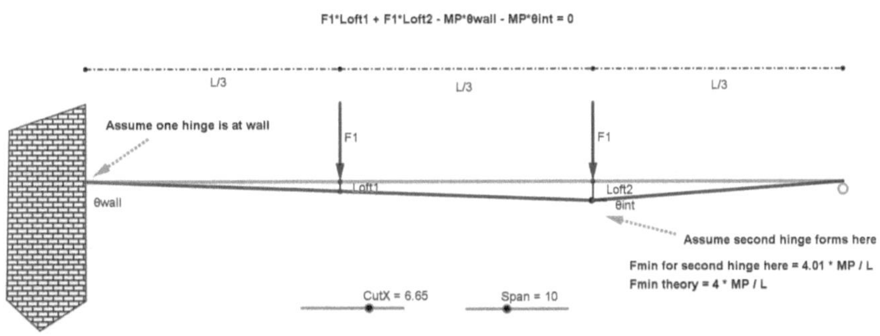

10.21
Minimum value of F1 found when the second hinge is at 2L/3

Having an error-free parametric drawing allows us to explore other failure scenarios. Notice in Figure 10.22 that the setup is similar, but the second load at 2 L /3 is some known factor of the first F1 load at L / 3. Here in this example, the second load is 1.5 times the first load. An arbitrary cut is taken and Equation 1.1 is used to find the Unknown F1. Here, we are not at the most dangerous, minimum load yet.

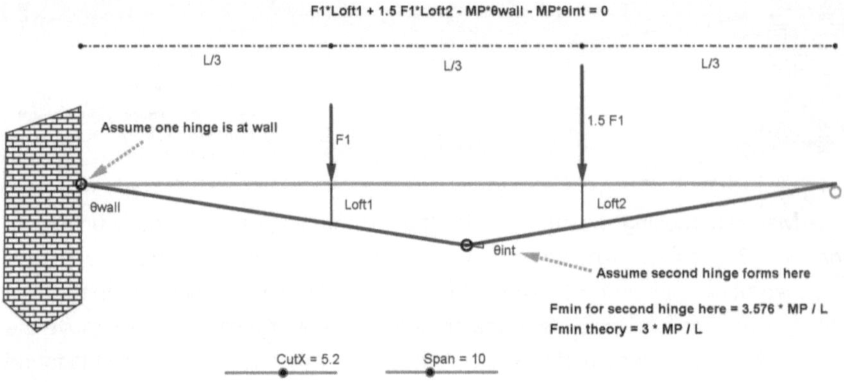

10.22
Known loads at known locations, but the second hinge location is not known

The ease of manipulating the parametric drawing allows for rapid exploration of various locations for the second hinge, even if the theoretical answer is not available to us. We fully recognize that the novice reader will not have the theoretical answer on hand, and as such, this method does not require such specialized knowledge. All that is needed is a quick geometric exploration and we soon find that placing that second hinge at the point of the 1.5 F1 load application gives the minimum value for F1. Figure 10.23 shows the minimum value of 3 MP/L which agrees with theory.

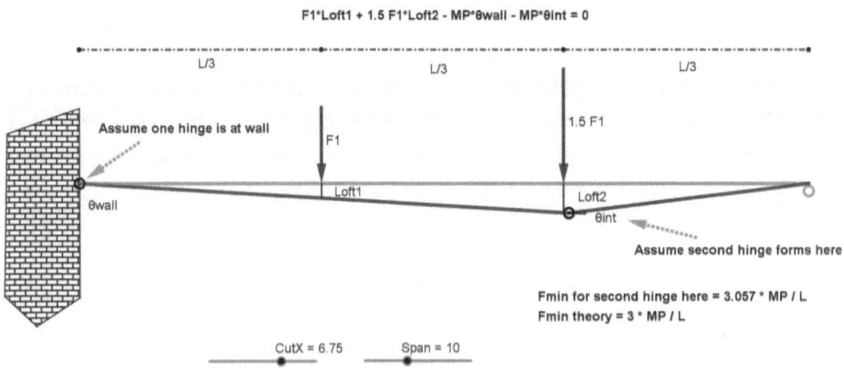

10.23
Known loads at known locations, move the second hinge location until minimum F1 is found

As a final example, consider the propped cantilever beam subjected to a uniformly distributed load, shown in Figure 10.24. We can assume that one hinge will form at the wall support, but now, it is not at all clear where the second hinge will form. Might it be at mid-span? Let's find out!

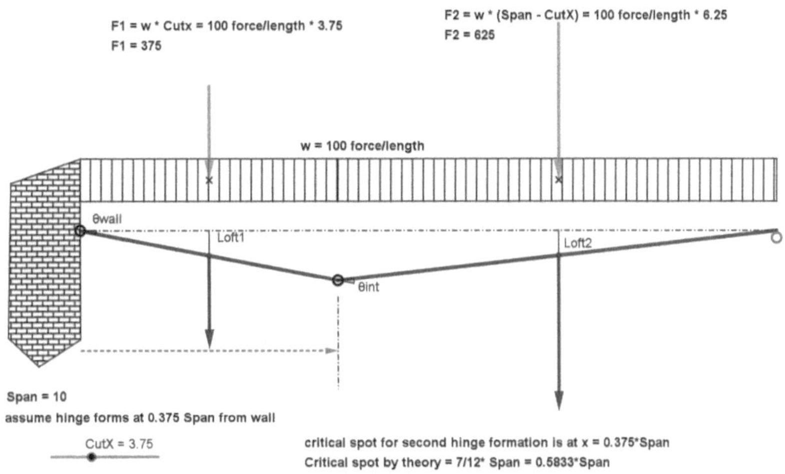

10.24
Propped cantilever, uniform load w: assume one hinge forms at the Left Wall: where is the second hinge?

Notice in Figure 10.24 that one of the fundamental Müller-Breslau Method rules must be used, namely that if the load is disbursed on either side of a kink, then a proportional concentrated load must be placed on either side of the kink. We can never use the total force Ftot = w*Span in such a problem, as a kink is necessary for the beam to collapse.

Perhaps it is somewhat surprising that the critical second hinge is not formed at the mid-span of the propped cantilever. Manual manipulation of the cut, with updates on the concentrated force on either side of the kink, rapidly finds that the minimum energy required to induce failure is when the cut is approximately at 0.58*span as shown in Figure 10.25.

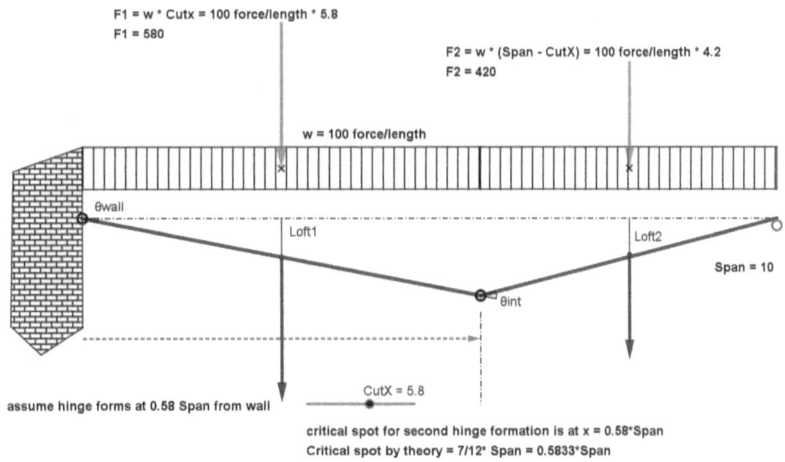

10.25
Getting close to the critical location for the second hinge

Since this is the Modern Müller-Breslau Method, the perturbation must be made small for the answer to asymptotically approach theory. Figure 10.26 shows the final solution for a small perturbation.

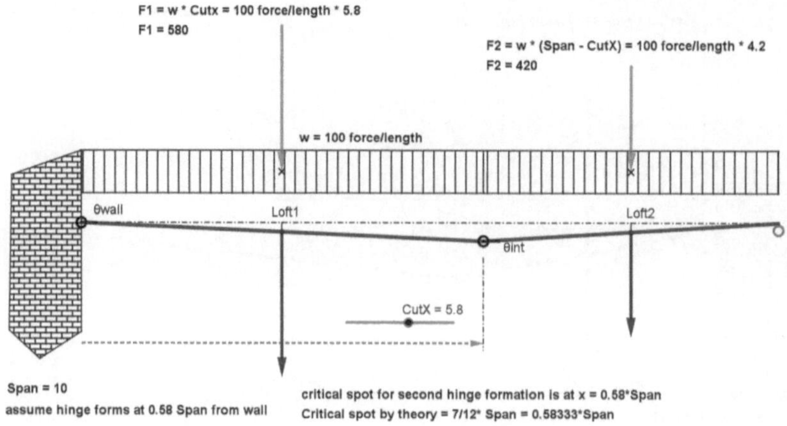

10.26
Final solution for the critical location of the second hinge

PLASTIC ANALYSIS OF STRUCTURES ÉTUDES

The following studies or études should be sketched by hand or in GeoGebra.

Étude 10.1) An indeterminate planar two-dimensional frame is shown in Figure 10.27. Some total load 3 F is applied to the top left moment carrying joint, and some total load 2 F is applied downward at the midspan of the beam. The frame is fixed at its left base and the pin supported at its right base. Note that the horizontal load and the vertical load are not independent of each other, they are both known factors of some load F. It is F that we will ultimately find. For this first étude, recreate the drawing of Figure 10.27, making sure that all the boundary conditions are understood.

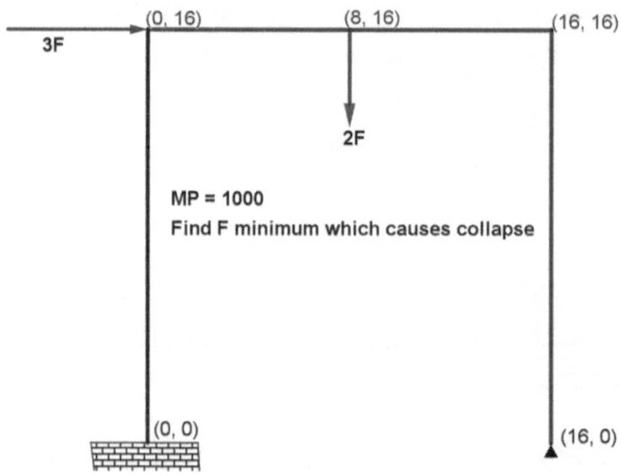

10.27
Étude 10.1 solution

Étude 10.2) Using Figure 10.27 as the original structure, imagine that the beam collapses due to three hinges being formed in it. How would this collapse mechanism appear? Label the angles as θ1, θ2, and θ3. Identify the lofts of the loads in this perturbed, failed, so-called "beam" mechanism. Figure 10.28 shows a solution. Note that the horizontal load has no loft in this mechanism.

Étude 10.3) Using Figure 10.27 as the original structure, imagine that the left column forms a hinge at its fixed base. Two more hinges are needed for collapse, they form at the top left and top right joints. How would this collapse mechanism appear? Label the angles as θ4, θ5, and θ6. Identify the lofts of the loads in this

10.28
Étude 10.2 solution

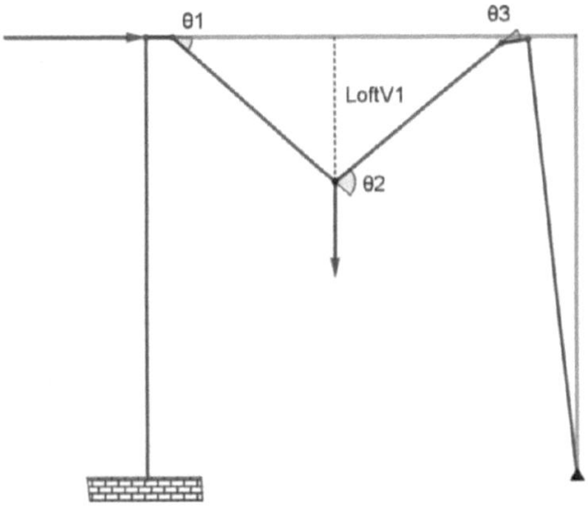

perturbed, failed, so-called "panel" mechanism. Figure 10.29 shows a solution. Note that the vertical load essentially has no loft in this mechanism as the perturbation gets very small.

10.29
Étude 10.3 solution

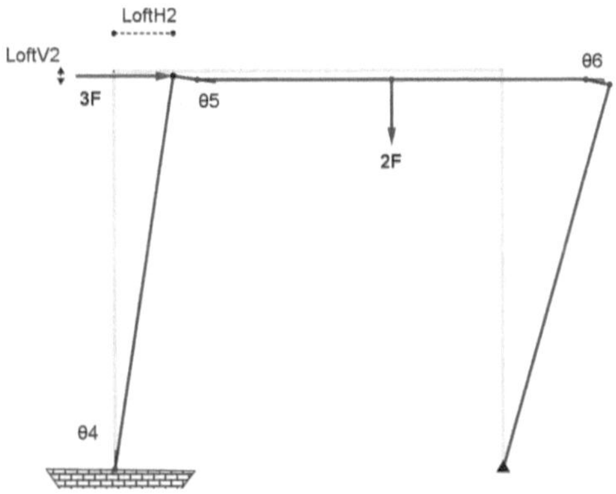

Étude 10.4) Using Figure 10.27 as the original structure, imagine that the left column forms a hinge at its fixed base. Two more hinges are needed for collapse, they form at the midspan of the beam and at the top right joint. How would this collapse mechanism appear? Label the angles as θ7, θ8, and θ9. Identify the lofts of the loads in this perturbed, failed, so-called "combined" mechanism. Figure 10.30 shows a solution.

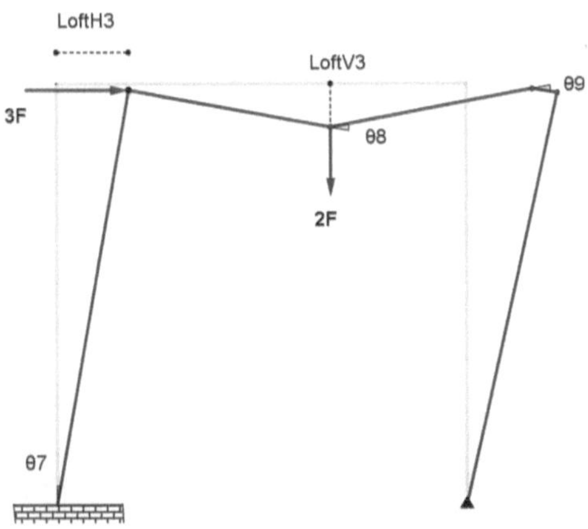

10.30
Étude 10.4 solution

Étude 10.5) Repeat étude 10.2, but now quantitatively find the F that causes a collapse in this "beam" mechanism. A solution is shown in Figure 10.31.

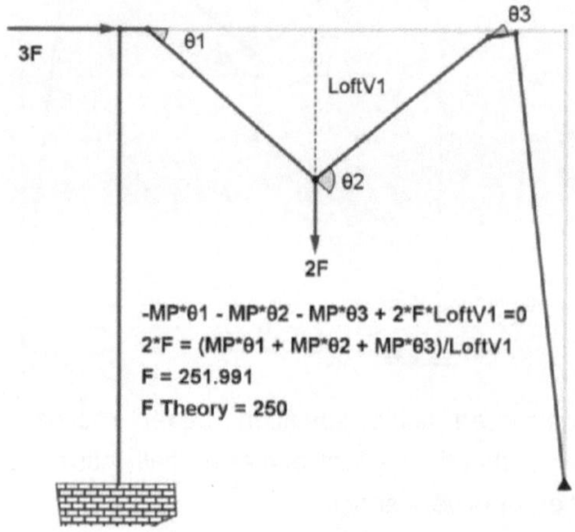

10.31
Étude 10.5 solution

Étude 10.6) Repeat étude 10.3, but now quantitatively find the F that causes a collapse in this "panel" mechanism. A solution is shown in Figure 10.32.

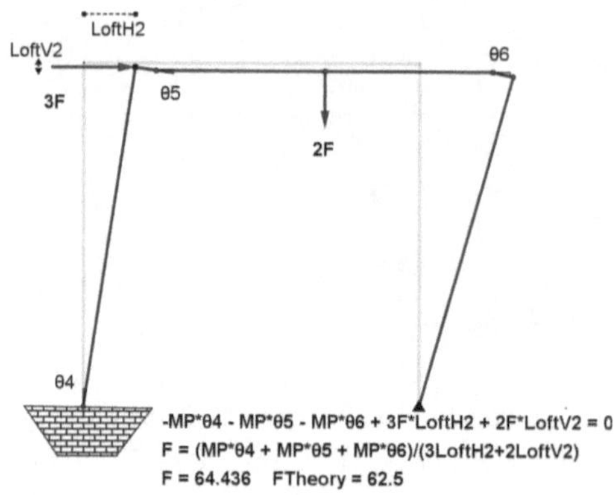

10.32
Étude 10.6 solution

Étude 10.7) Repeat étude 10.4, but now quantitatively find the F that causes a collapse in this "combined" mechanism. A solution is shown in Figure 10.33.

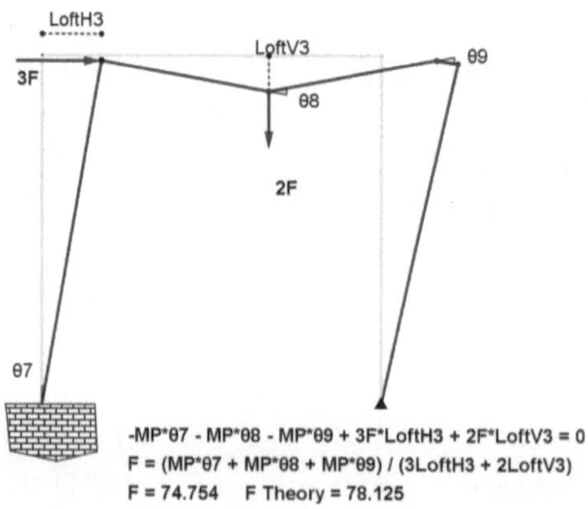

10.33
Étude 10.7 solution

Étude 10.8) Using all the drawings of these Chapter 10 études, determine what is the minimum load that actually causes a collapse of the frame being studied.

Answer: The so-called "panel" mechanism, with F minimum of approximately 62.5 units of force as F. Thus, 3F = 187.5 units of force and 2F = 125 units of force.

Index